AI in Marketing, Sales and Service

Peter Gentsch

AI in Marketing, Sales and Service

How Marketers without a Data Science Degree can use AI, Big Data and Bots

Peter Gentsch
Frankfurt, Germany

ISBN 978-3-319-89956-5   ISBN 978-3-319-89957-2   (eBook)
https://doi.org/10.1007/978-3-319-89957-2

Library of Congress Control Number: 2018951046

Cover illustration: Andrey Suslov/iStock/Getty
Cover design by Tom Howey

This Palgrave Macmillan imprint is published by the registered company Springer Nature Switzerland AG
The registered company address is: Gewerbestrasse 11, 6330 Cham, Switzerland

Contents

Notes on Contributors

Alex Dogariu has over 10 years of experience in customer management, corporate strategy and disruptive technologies (e.g. artificial intelligence, RPA, blockchain) in e-commerce, banking services and automotive OEMs. Alex began his career at Accenture, driving CRM and sales strategy innovations. He then moved on to be managing director at logicsale AG, revolutionizing e-commerce through dynamic repricing. In 2015, he joined Mercedes-Benz Consulting, leading the customer management strategy and innovation department. He was recently awarded twice the 1st place in the Best of Consulting competition hosted by *WirtschaftsWoche* in the categories Digitization as well as Sales and Marketing.

Klaus Eck is a blogger, speaker, author and founder of the content marketing agency d.Tales.

Prof. Dr. rer. pol. Nils Hafner is an international expert in building consistently profitable customer relations. He is professor for customer relationship management at the Lucerne University of Applied Sciences and Arts and heads a program for customer relations management.

Prof. Dr. Hafner studied economics, psychology, philosophy and modern history in Kiel and Rostock (Germany). He earned his Ph.D. in innovation management/marketing with a dissertation on KPIs of call center services. After his engagement as a practice leader CRM in one of the largest business consulting firms, he established from 2002 to 2006 the first CRM Master program in the German-speaking countries.

At present, he advises the management of medium-sized and major enterprises in Germany, Switzerland and Europe in matters of CRM. In his blog

"Hafner on CRM", he is trying to emphasize the informative, delightful, awkward, tragic and funny aspects of the subject. Since 2006, he publishes the "Top 5 CRM Trends of the Year" and speaks about these trends in over 80 Speeches per year for international top companies.

Bruno Kollhorst works as Head of advertising and HR-marketing at Techniker Krankenkasse (TK), Germanys biggest public health insurance company. He is also member of the Social Media Expert Board at BVDW. The media and marketing-specialist works also as lecturer at University of Applied Sciences in Lübeck and is a freelance author. Beneath advertising, content marketing and its digitalization, he is also an expert in the sectors brand cooperation and games/e-sports.

Jens Scholz studied mathematics at the TU Chemnitz with specialization in statistics. After this, he worked as managing director of die WDI media agentur GmbH. He is one of the founders of the prudsys AG. Since 2003 he was responsible for marketing and later sales at prudsys. Since 2006 he is the CEO of the company.

Andreas Schwabe in his role as Managing Director of Blackwood Seven Germany, he revolutionizes media planning through artificial intelligence and machine learning. With a specifically developed platform, the software company calculates for each customer the "Media Affect Formula", which enables an attribution of all online channels such as Search, YouTube and Facebook along with offline such as TV, radio broadcast, print and OOH. This simulates the ideal media mix for the customers. Blackwood Seven has 175 employees in Munich, Copenhagen, Barcelona, New York and Los Angeles.

Dr. Michael Thess studied mathematics in Chemnitz und St. Petersburg. He specialized in numerical analysis and received the Ph.D. at the TU Chemnitz. As one of the founders of the prudsys AG, he was responsible for research and development. Since 2017 he manages the Signal Cruncher GmbH, a daughter company of prudsys.

Dr. Thomas Wilde is an entrepreneur and lecturer at LMU Munich. His area of expertise lies in digital transformation, especially in software solutions for marketing and service in social media, e-commerce, messaging platforms and communities.

Prior to that, he worked as an entrepreneur, consultant and manager in strategic business development. He studied economics and did his doctor's degree in business informatics and new media at the Ludwig-Maximilian University in Munich.

List of Figures

List of Tables

Part I

AI 101

1

AI Eats the World

Artificial intelligence (AI) has catered for an immense leap in development in business practice. AI is also increasingly addressing administrative, dispositive and planning processes in marketing, sales and management on the way to the holistic algorithmic enterprise. This introductory chapter deals with the motivation for and background behind the book: It is meant to build a bridge from AI technology and methodology to clear business scenarios and added values. It is to be considered as a transmission belt that translates the informatics into business language in the spirit of potentials and limitations. At the same time, technologies and methods in the scope of the chapters on the basics are explained in such a way that they are accessible even without having studied informatics—the book is regarded as a book for business practice.

1.1 AI and the Fourth Industrial Revolution

If big data is the new oil, analytics is the combustion engine (Gartner 2015).

Data is only of benefit to business if it is used accordingly and capitalised. Analytics and AI increasingly enable the smart use of data and the associated automation and optimisation of functions and processes to gain advantages in efficiency and competition.

AI is not another industrial revolution. This is a new step on the path of the universe. The last time we had a step of that significance was 3.5 billion years ago with the invention of life.

© The Author(s) 2019
P. Gentsch, *AI in Marketing, Sales and Service*,
https://doi.org/10.1007/978-3-319-89957-2_1

In recent years, AI has catered for an immense leap in development in business practice. Whilst the optimisation and automation of production and logistics processes are focussed on in particular in the scope of Industry 4.0, AI increasingly also addresses administrative, dispositive and planning processes in marketing, sales and management on the path towards the holistic algorithmic enterprise.

AI as a possible mantra of the massive disruption of business models and the entering of fundamental new markets is asserting itself more and more. There are already many cross-sectoral use cases that give proof of the innovation and design potential of the core technology of the twenty first century. Decision-makers of all industrial nations and sectors are agreed. Yet there is a lack of a holistic evaluation and process model for the many postulated potentials to also be made use of. This book proposes an appropriate design and optimisation approach.

Equally, there is an immense potential for change and design for our society. Former US President Obama declared the training of data scientists a priority of the US education system in his keynote address on big data. Even in Germany, there are already the first data science studies to ensure the training of young talents. In spite of that, the "war of talents" is still on the rampage as the pool of staff is still very limited, with the demand remaining high in the long term.

Furthermore, digital data and algorithms facilitate totally new business processes and models. The methods applied range from simple hands-on analytics with small data down to advanced analytics with big data such as AI.

At present, there are a great many informatics-related explanations by experts on AI. In equal measure, there is a wide number of popular scientific publications and discussions by the general public. What is missing is the bridging of the gap from AI technology and methodology to clear business scenarios and added values. IBM is currently roving around from company to company with Watson, but besides the teaser level, the question still remains open about the clear business application. This book bridges the gap between AI technology and methodology and the business use and business case for various industries. On the basis of a business AI reference model, various application scenarios and best practices are presented and discussed.

After the great technological evolutionary steps of the Internet, mobiles and the Internet of Things, big data and AI are now stepping up to be the greatest ever evolutionary step. The industrial revolution enabled us to get rid of the limitations of physical work like these innovations enable us to overcome intellectual and creative limitations. We are thus in one of the most thrilling phases of humanity in which digital innovations fundamentally change the economy and society.

1.2 AI Development: Hyper, Hyper…

If we take a look at business articles of the past 20 years, we notice that every year, there is always speak of the introduction of "constantly increasing dynamisation" or "shorter innovation and product cycles"—similar to the washing powder that washes whiter every year. It is thus understandable that with the much-quoted speed of digitalisation, a certain degree of immunity against the subject has crept into one person or the other. The fact that we have actually been exposed to a non-existing dynamic is illustrated by Fig. 1.1: On the historic time axis, the rapid peed of the "digital hyper innovation" with the concurrently increasing effect on companies, markets and society becomes clear. This becomes particularly clear with the subject of AI.

The much-quoted example of the AI system AlphaGo, which defeated the Korean world champion in "Go" (the world's oldest board game) at the beginning of 2016 is an impressive example of the rapid speed of development, especially when we look at the further developments and successes in 2017.

The game began at the beginning of 1996 when the AI system "Deep Blue" by IBM defeated the reigning world champion in chess, Kasparow. Celebrated in public as one of the breakthroughs in AI, the enthusiasm among AI experts was contained. After all, in the spirit of machine

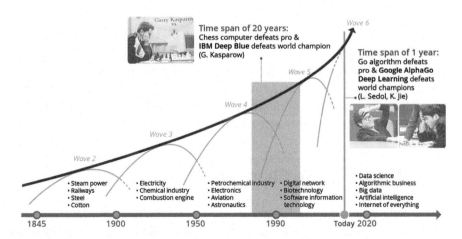

Fig. 1.1 The speed of digital hyper innovation

learning, the system had quite mechanically and, in fact, not very intelligently, discovered success patterns in thousands of chess games and then simply applied these in real time faster than a human could ever do. Instead, the experts challenged the AI system to beat the world champion in the board game "Go". This would then have earned the attribute "intelligent", as Go is far more complex than chess and in addition, demands a high degree of creativity and intuition. Well-known experts predicted a period of development of about 100 years for this new milestone in AI. Yet as early as March 2016, the company DeepMind (now a part of Google) succeeded in defeating the reigning Go world champion with AI. At the beginning of 2017, the company brought out a new version of AlphaGo out with Master, which has not only beaten 60 well-experienced Go players, but had also defeated the first version of the system that had been highly celebrated only one year prior. And there's more: In October 2017 came Zero as the latest version, which not only defeated AlphaGo but also its previous version. The exciting aspect about Zero is that, on the one hand, it got by with a significantly leaner IT infrastructure, on the other hand, in contrast to its previous version, it was not fed any decided experience input from previously played games. The system learned how to learn. And in addition to that, with fully new moves that the human race had never made in thousands of years. This proactive, increasingly autonomous acting makes AI so interesting for business. As a country that sees itself as the digital leader, this "digital hyper innovation" should be regarded as the source of inspiration for business and society and be used, instead of being understood and repudiated as a stereotype as a danger and job killer.

The example of digital hyper innovation shows vividly what a nonlinear trend means and what developments we can look forward to or be prepared for in 2018. In order to emphasise this exponentiality once again with the board game metaphor: If we were to take the famous rice grain experiment by the Indian king Sheram as an analogy, which is frequently used to explain the underestimation of exponential development, the rice grain of technological development has only just arrived at the sixth field of the chess board.

1.3   AI as a Game Changer

In the early phases of the industrial revolutions, technological innovations replaced or relieved human muscle power. In the era of AI, our intellectual powers are now being simulated, multiplied and partially even substituted by digitalisation and AI. This results in fully new scaling and multiplication effects for companies and economies.

Companies are developing increasingly strongly towards algorithmic enterprises in the digital ecosystems. And it is not about a technocratic or mechanistic understanding of algorithms, but about the design and optimisation of the digital and analytical value added chain to achieve sustainable competitive advantages. Smart computer systems, on the one hand, can support decision-making processes in real time, but furthermore, big data and AI are capable of making decisions that today already exceed the quality of human decisions.

The evolution towards the algorithmic enterprise in the spirit of the data- and analytics-driven design of business processes and models directly correlates with the development of the Internet. However, we will have to progressively bid farewell to the narrow paradigm of usage of the user sitting in front of the computer accessing a website. "Mobile" has already changed digital business significantly. Thanks to the development of the IoT, all devices and equipment are progressively becoming smart and proactively communicate with each other. Conversational interfaces will equally change human-to-machine communication dramatically—from the use of a text-based Internet browser down to natural language dialogue with everybody and everything (Internet of Everything).

Machines are increasingly creating new scopes for development and possibilities. The collection, preparation and analysis of large amounts of data eats up time and resources. The work that many human workers used to perform in companies and agencies is now automated by algorithms. Thanks to new algorithmics, these processes can be automated so that employees have more time for the interpretation and implementation of the analytical results.

In addition, it is impossible for humans to tap the 70 trillion data points available on the Internet or unstructured interconnectedness of companies and economic actors without suitable tools. AI can, for example, automate the process of customer acquisition and the observation of competition so that the employees can concentrate on contacting identified new customers and on deriving competitive strategies.

Recommendations and standard operation procedures based on AI and automated evaluation are often eyed critically by companies. It surely feels strange at the beginning to follow these automated recommendations that are created from algorithms and not from internal corporate consideration. However, the results show that it is worthwhile because we are already surrounded by these algorithms today. The "big players" (GAFA = Google,

Apple, Facebook, Amazon) are mainly to solely relying on algorithms that are classified in the category "artificial intelligence" for good reason. The advantage: These recommendations are free of subjective influences They are topical, fast and take all available factors into consideration.

Even at this stage, the various successful use and business cases for the AI-driven optimisation and design of business processes and models can be illustrated (Chapter 5). What they all have in common is the great change and disruption potential The widespread mantra in the digital economy of "software eats the world" can now be brought to a head as "AI & algorithmics eat the world".

1.4   AI for Business Practice

Literature on the subject of big data and AI is frequently very technical and informatics-focused. This book sees itself as a transmission belt that translates the language of business in the spirit of potentials and limitations. At the same time, the technologies and methods do not remain to be a black box. They are explained in the scope of the chapters on the basics in such a way that they are accessible even without having studied informatics.

In addition, the frequently existing lack of imagination between the potentials of big data, business intelligence and AI and the successful application thereof in business practice is closed by various best practice examples. The relevance and pressure to act in this area do happen to be repeatedly postulated, yet there is a lack of a systematic reference frame and a contextualisation and process model on algorithmic business. This book would like to close that roadmap and implementation gap.

The discussion on the subjects is very industry-oriented, especially in Germany. Industry 4.0, robotics and the IoT are the dominating topics. The so-called customer facing functions and processes in the fields of marketing, sales and service play a subordinate role in this. As the lever for achieving competitive advantages and increasing profitability is particularly high in these functions, this book has made it its business to highlight these areas in more detail and to illustrate the outstanding potential by numerous best practices:

- How can customer and market potentials be automatically identified and profiled?
- How can media planning be automated and optimised on the basis of AI?

- How can product recommendations and pricing be automatically derived and controlled?
- How can processes be controlled and coordinated smartly by AI?
- How can the right content be automatically generated on the basis of AI?
- How can customer communication in service and marketing be optimised and automated to increase customer satisfaction?
- How can bots and digital assistants make the communication between companies and consumers more efficient and more smart?
- How can the customer journey optimisation be optimised and automated on the basis of algorithmics and AI?
- What significance do algorithmics and AI have for Conversational Commerce?
- How can modern market research by optimised intelligently?

Various best practice examples answer these questions and demonstrate the current and future business potential of big data, algorithmics and AI (Chapter 5 AI Best Practices).

Reference

Gartner. (2015). *Gartner Reveals Top Predictions for IT Organizations and Users for 2016 and Beyond.* http://www.gartner.com/newsroom/id/3143718. Accessed 5 Jan 2017.

2

A Bluffer's Guide to AI, Algorithmics and Big Data

2.1 Big Data—More Than "Big"

A few years ago, the keyword big data resounded throughout the land. What is meant is the emergence and the analysis of huge amounts of data that is generated by the spreading of the Internet, social media, the increasing number of built-in sensors and the Internet of Things, etc.

The phenomenon of large amounts of data is not new. Customer and credit card sensors at the point of sale, product identification via barcodes or RFID as well as the GPS positioning system have been producing large amounts of data for a long time. Likewise, the analysis of unstructured data, in the shape of business reports, e-mails, web form free texts or customer surveys, for example, is frequently part of internal analyses. Yet, what is new about the amounts of data falling under the term "big data" that has attracted so much attention recently? Of course, the amount of data available through the Internet of Things (Industry 4.0), through mobile devices and social media has increased immensely (Fig. 2.1).

A decisive factor is, however, that due to the increasing orientation of company IT systems towards the end customer and the digitalisation of business processes, the number of customer-oriented points of contact that can be used for both generating data and systematically controlling communication has increased. Added to this is the high speed at which the corresponding data is collected, processed and used. New AI approaches raise the analytical value creation to a new level of quality.

© The Author(s) 2019
P. Gentsch, *AI in Marketing, Sales and Service*,
https://doi.org/10.1007/978-3-319-89957-2_2

Fig. 2.1 Big data layer (Gentsch)

2.1.1 Big Data—What Is Not New

The approach of gaining insights from data for marketing purposes is nothing new. Database marketing or analytical CRM has been around for more than 20 years. The phenomenon of large amounts of data is equally nothing new: Point of sale, customer and credit cards or web servers have long been producing large amounts of data. Equally, the analysis of unstructured data in the shape of emails, web form free texts or customer surveys, for example, frequently form a part of marketing and research.

2.1.2 Big Data—What Is New

It goes without saying that the amount of data has increased immensely thanks to the Internet of Things, mobiles and social media—yet this is rather a gradual argument. The decisive factor is that thanks to the possibilities of IT and the digitalisation of business processes, customer-oriented points of contact for both generating data and for systematically controlling communication have increased. Added to this is the high speed at which the corresponding data is collected, processed and used. Equally, data mining methods of deep learning and semantic analytics raise the analytical value creation to a new level of quality.

2.1.3 Definition of Big Data

As there are various definitions of big data, one of the most common ones will be used here:

> "Big data" refers to datasets whose size is beyond the ability of typical database software tools to capture, store, manage, and analyse. (Manyika et al. 2011)

Following this definition, big data has been around ever since electronic data processing. Centuries ago, mainframes were the answer to ever-increasing amounts of data and the PCs of today have more storage space and processing power than those mainframes of back then.

In the infographic of IBM, big data is frequently described using the four Vs: What they mean are the following dimensions of big data

- Volume: This describes the amount of incoming data that is to be stored and analysed. The point when an amount of data is actually declared as big data as described above depends on the available systems. Companies are still facing the challenge of storing and analysing incoming amounts of data both efficiently and effectively. In recent years, various technologies such as distributed systems have become established for these purposes.
- Velocity: This describes two aspects: On the one hand, data is generated at a very high speed and, on the other hand, systems must be able to store, process and analyse these amounts of data promptly. These challenges are tackled both by hardware with the help of in-memory technologies, for example,[1] as well as by software, with the help of adapted algorithms and massive parallelisation.
- Variety: The great variety of data of the world of big data confronts systems with the task of no longer only processing with structured data from tables but also with semi- and unstructured data from continuous texts, images or videos, which make up as much as 85% of the amounts of data. Especially in the field of social media, a plethora of unstructured data is accumulated, whose semantics can be collected with the help of AI technologies.
- Veracity: Whereas the three dimensions described here can be mastered by companies today with the help of suitable technologies, methods and the use of sufficient means, there is one challenge that has not yet been solved to the same extent. Veracity means the terms of trustworthiness, truthfulness and meaningfulness of big data. It is thus a matter of not all stored data is trustworthy and this should not be analysed. Examples of this are manipulated sensors in the IoT, phishing mails or, ever since the last presidency election in the USA, also fake news.

A wide number of methods of AI is used for the evaluation and analysis of big data. In the following subchapter, the synergy effects of big data and AI are explained.

2.2   Algorithms—The New Marketers?

Data—whether small, big or smart—does not yield added value per se. It is algorithms, whether simple predefined mechanisms or self-learning systems, that can create values from the data. In contrast to big data, it is the algorithms that have a real value. Dynamic algorithms are taking centre stage in future digital business. Algorithms will thus become increasingly important for analysing substantially increasing amounts of data. This chapter is dedicated to the "power" and increasing significance and relevance of algorithms, undertakes an attempt at a definition, studies success factors and drivers of AI and further takes a glance at the historical development of artificial intelligence from the first works until today. Finally, the key methods and technologies for the AI business framework will be presented and explained.

In times when the mass of data doubles about every two years, algorithms are becoming more and more important for analysing this data. Whilst data is called the gold of the digital era, it is the possibilities of analysing this data to become usable results that generate the effective value. Complex algorithms are thus frequently called the driving force of the digital world. Applied with the right business model, they open up new opportunities and increasing competitive advantages.

The potential emanating from big data was recognised at an early stage and it still remains topical. However, the new challenges no longer lie solely in the collection storage and analysis of this data. The next step that is currently causing many companies a headache is the question of its benefit. That is precisely the task of algorithmic business. The point here is to take the next step towards a fully automated company. This is to be achieved by the use of smart algorithms that not only serve the purpose of evaluating and analysing data, but which also derive independent actions resulting from the analyses. These fully autonomous mechanisms that run in the background are contributing ever larger shares in the value creation of companies. Similar to the intelligence and algorithmics of self-driving cars, these technologies can successively assume the control and autonomy of companies.

The term algorithm was typically always associated with the subjects of mathematics and informatics. Today, the term algorithm is also strongly boosted by public discourse. The rather "innocent, somewhat boringly dustily connotated" term has now become a phenomenon that, against the background of the fourth industrial revolution and the threatening front of the substitution of jobs, is being discussed critically in public.

The term algorithm is also frequently used as a "fog bomb" when organisations either did not want to or could not explain to the consumer why which action was chosen. In fact, it was explained by saying that something very complex was happening in the computer. Consequently, the term algorithm is used on the one hand secretively and on the other hand, as a substitute when it comes to rewriting would-be complex circumstances or to explain to oneself the "miracle" of the digital present age. This is why it is hardly surprising that the term is unsettling in public discussion and makes it difficult for beginners to actually estimate the potential and risk. The "power of the algorithm" is perceived by some with awe; others, in contrast, are scared of it, whereby these strands sometimes merge when the algorithm is described as an "inscrutable, oracle-like" power.

The subject of algorithmics is also frequently associated with the topic of algorithmic personalisation. Be it the initially chronologically produced and today personally subscribable news feed on Facebook, the personalised Google search launched in 2009 or the likes of suggestions by Netflix and Spotify—they all work with algorithms that serve the purpose of personalising the contents played out. The starting point is usually a collected customer profile, which is used by the corresponding institutions to issue tailor-made recommendations to the user. This ranges from recommended purchases (e.g. Amazon) down to the recommendation of potential partners (e.g. Parship). Algorithms have many far-reaching application scenarios and implications as will be shown in the following chapter.

2.3 The Power of Algorithms

Algorithms are meant to optimise or even re-create operational functions and value added chains by way of accuracy, sped and automation. With that, the question is posed as to how algorithms are to be developed and fed. And in turn, it has less to do with the software-technical programming capacity, but in fact the underlying knowledge base. Figure 2.2 shows the correlation between algorithmics and artificial intelligence. The correlation is determined by the complexity and degree of structuring of the underlying tasks.

Simple algorithms are defined and executed via rules. These can be, for example, event-driven process chains (EPCs). The event "customer A calls the call centre" can trigger the call to be passed on to particularly experienced staff. Such workflows are driven by previously defined rules.

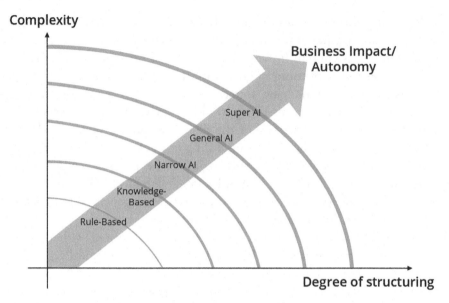

Fig. 2.2   Correlation of algorithmics and artificial intelligence (Gentsch)

Marketing automation solutions also allow for the defining of such rules for the systematic automation of customer communication (for example the rule for lead nurturing or drip campaigns).

However, it is difficult to solve more complex and less structured tasks by way of predefined rules. This is where knowledge-based systems can help. For example, a complex, previously unknown problem a customer has can be solved by a so-called case-based reasoning system. The algorithm operationalises the enquiry (definition of a so-called case) and looks for similar, already solved problems (cases) in a knowledge database. Then, by way of an analogy conclusion, a solution is derived for the new, still unknown problem.

Methods of artificial intelligence can be applied for even more complex, unstructured tasks. At present, the AI applications belong to the so-called narrow intelligence. An AI system is developed for a certain domain. This could be, for example, a deep learning algorithm that automatically predicts and profiles matching leads on the basis of big data on the Internet (Sect. 5.1 "Sales and Marketing Reloaded").

AI applications of general intelligence (human intelligence level) and super intelligence (singularity) do not exist at present. The challenge here is in the necessary transfer performance between different domains. These systems could then proactively and dynamically develop and execute their own algorithm solutions depending on the context. In Sect. 3.4 ("AI Maturity Model") companies are described as an example in the dimensions strategy, people/orga, data and analytics that have the necessary algorithmic maturity level for this.

Overall, the necessary autonomy and dynamics of algorithms is increasing with the increasing complexity and decreasing degree of structure of the task. This also applies to the business impact in the spirit of competitive relevance of the algorithm solutions.

2.4 AI the Eternal Talent Is Growing Up

The subject of AI is nothing new—it has been discussed since the 1960s. The great breakthrough in the business world has failed to appear, but for a few exceptions. Thanks to the immensely increased computing power, the methods can now be massively parallelised and intensified. Innovative deep learning and predictive analytics methods paired with big data technology facilitate a quantum leap of AI potential benefits for business applications and problems. In the last ten years, the breakthrough with regard to the applicability in business practice has succeeded due to this further development. At present, the discussion is, on the one hand, shaped by hardly realistic science fiction scenarios that postulate computers taking over mankind. On the other hand, there is a strongly informatics-/technology-laden discourse. In addition to that, there are singular popular science publications as well as articles in the daily press. The latter adhere to the exemplary level without holistic context. A systematic overview of the AI relevant for business, a reference model for classification for the respective business functions and problems, a maturity model for the classification and evaluation of the respective phases and a process model including an economic cost-benefit analysis are all lacking.

2.4.1 AI—An Attempt at a Definition

Hardly any other field of informatics triggers emotions as frequently as the field called "artificial intelligence" does. The term firstly reminds us of intelligent human robots as known from science fiction novels and films. The questions are quickly posed as to: "Will machines be intelligent one day?" or "will machines be able to think like humans?" There are countless attempts at defining the term artificial intelligence that, depending on the expert and historic origin, have a different focus and a different faceting.

Yet, before we try to occupy ourselves with "artificial intelligence" we should first define "intelligence". There has not been a holistic definition of it yet, as intelligence exists on various levels and there is no consensus as to how it is to be differentiated. However, a core statement can be recognised in many cases. Intelligence is the "ability [of a human] of abstract and reasonable thinking and to derive purposeful actions from it" (as per Duden 2016).

In essence, it is "a general mental ability that, among others, covers recognising rules and reasons, abstract thinking, learning from experience, developing complex ideas, planning and solving problems" (Klug 2016). Artificial intelligence must therefore reproduce the named aspects of human behaviour, in order to be able to act "human" in this way, without being human. This includes traits and skills such as solving problems, explaining, learning, understanding speech as well as a human's flexible reactions.

As it is not possible to find the absolutely true definition of artificial intelligence, the following definition by Elaine Rich seems to be the one best suited for this book:

Artificial Intelligence is the study of how to make computers do things at which, at the moment, people are better. (Rich 2009)

This expresses that AI is always relative as a kind of competition between man and machine over time and in its distinctness and performance. Just like Deep Blue defeating Kasparow in 1996 was celebrated, it was the Jeopardy victory in 2011 and the victory of AI over the Korean world champion in Go in 2016.

2.4.2 Historical Development of AI

The history of artificial intelligence can be divided into various phases. In the scope of this book, a short overview will be given of the individual stages of development of artificial intelligence from the beginning in the 1950s to today (Fig. 2.3).

2.4.2.1 First Works in the Field of Artificial Intelligence (1943–1955)

In 1943, the Americans Warren McCulloch (1898–1969) and Walter Pitts (1923–1969) published the first work dedicated to the field of AI

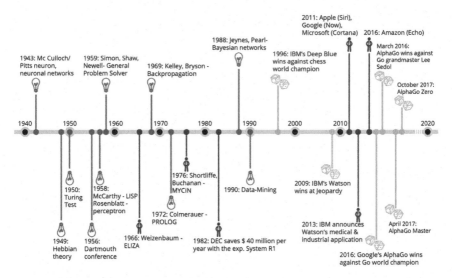

Fig. 2.3 Historical development of AI

(Russell and Norvig 2012). Based on knowledge from the disciplines neurology, mathematics and programming theory, they presented the so-called McCulloch-Pitts Neuron. They describe for the first time as an example the structure of artificial neuronal networks, the set-up and structure of which are based on the human brain. At the same time, individual neurons can adopt various states ("on" or "off"). By combining the neurons and their interactions, information can be stored, changed and computed. In addition, McCulloch and Pitts prophesied that such network structures can also be adaptive with the right configuration (Russell and Norvig 2012). The concepts presented back then were promising, yet an implementation on a grand scale would not have been technically possible at that time due to the lack of IT infrastructures.

The most significant articles were those by Alan Turing (1912–1954), who had already given speeches on AI at the London Mathematical Society in as early as 1947 and, in 1950, he published his visions in the article "Computing Machinery and Intelligence" (Russell and Norvig 2012). In the paper that was published in the philosophical journal "Mind", Turing asked the crucial question of AI: "Can Machines Think". In addition, in the article, he presented his ideas according to the Turing test named after him, machine learning, genetic algorithms and reinforcement learning.

2.4.2.2 Early Enthusiasm and Speedy Disillusion (1952–1969)

The term "artificial intelligence" was first spoken of at a conference held at Dartmouth College in Hanover in the US State of New Hampshire in 1956. At the invitation of John McCarthy (1927–2011), leading researchers from America came together there. In the two-month workshop, subjects such as neuronal networks, automatic computers and the attempt to teach speech to computers were to be handled. At this workshop, there were in fact no new breakthroughs, yet the conference is still considered a milestone because the most important pioneers of the development of AI of that time met up and established the science of artificial intelligence (Russell and Norvig 2012).

The Turing test is a test to establish human-like intelligence in a machine. To this end, a person communicates via text chat with two people unknown to him, of which one is a human and the other a machine. Both try to convince the interrogator that they are humans. The test is deemed passed when the computer succeeds in not standing out as a computer to his human opposite in more than 30% of a series of short conversations, and if the human cannot differentiate between man and machine with certainty. There has not been a program to this day that has passed the Turing test indisputably.

In the years that followed, great enthusiasm about the future developments and successes of artificial intelligence proliferated. This is what the later winner of the Turing Award and Nobel Prize in Economics, Herbert A. Simon (1916–2001), postulated in 1958.

Within the next ten years, a computer will become the chess world champion and within the next ten years, an important new mathematical theory will be discovered and proven.

2.4.2.3 Knowledge-Based Systems as the Key to Commercial Success (1969–1979)

The methods used up until now, also called "weak methods" where search algorithms combine elementary sub-steps to get to the solution to the problem, were not able to solve any complex problems. For this reason, the approach was adapted in the 1970s. Instead of programs whose approaches can be applied to a large number of problems, methods were developed that use area-specific knowledge and methods of the respective specialist field. For this purpose, complex rules and standards were formed within which the program arrives at the solution. The so-called expert systems were meant to bring about success especially in the fields of speech recognition, automatic translation and medicine (Russell and Norvig 2012).

2.4.2.4 The Return to Neuronal Networks and the Ascension of AI to Science (1986 to Today)

In the middle of the "AI winter", the psychologists David Rumelhart and James McClelland revived in an article interest in the back propagation algorithm that had already been published in 1969. This could be applied to various problems of informatics and psychology. This caused research into neuronal networks to be revived and two key branches of AI research arose:

- The symbolic, logical approach that pursues the top-down approach and systematically links expert knowledge, as well as codifies with the help of complex rules and standards, to be able to make conclusions (Russell and Norvig 2012), and
- The neuronal AI, whose methods are geared to the way the human brain works. This approach is responsible for the current euphoria around AI.

Neuro-informatics, which deals with the part of AI with the same name, has been able to make notable progress in the last two decades with the help of other scientific disciplines such as psychology, neurology, linguistics and cognitive sciences and has thus attracted attention to itself from the business world, politics and society. This is why the field of AI research is no longer considered in isolation from other disciplines, but understood as a combination of various fields of research.

2.4.2.5 Intelligent Agents Are Becoming a Normality (1995 to Today)

Until now, neither the united exertions of different scientific disciplines nor huge amounts of funding for projects such as the Human Brain Project with funding of 1.2 billion EUR, have been able to lead to the development of artificial intelligence equal to a human. A machine thinking in such a way would be a so-called general artificial intelligence (also called AGI or strong AI), i.e. a mechanism that would be able to perform any intellectual tasks like they would equally be performed by a human or even better. Whilst AI research in this area is still far from its goal, at present, a great number of systems that are classified in the area "artificial narrow intelligence" (ANI) are being developed and have been used for decades. Systems on the Internet are known to most people under the name of bot.

These computer programs are capable of acting autonomously within a defined environment. Whilst pioneering experts such as #MinsAI and McCarthy criticise the fact that there is only little commercial interest in the development of an AGI or a human-level AI (HLAI), the public sector develops systems in many areas that can be classified under narrow AI. Intelligent agents are most frequently encountered on the Internet. There, they act as parts of search engines, crawlers or recommendation systems. The levels of complexity of intelligent agents vary from simple scripts to sophisticated chatbots that simulate human-like intelligence.

The number of scientific publications doubles every nine years. The growth rates of the AI publications from 1960 to 1995 in contrast lie at more than 100% every five years, and between 1995 and 2010, they were still more than 50% every five years.

2.4.3 Why AI Is Not Really Intelligent—And Why That Does Not Matter Either

Despite the great AI successes of recent years, we are still in an era of very formal, machine AI. Figure 2.4 shows that the underlying methods and technologies have not fundamentally changed since the 1950s/1960s to today. However, due to the increased amounts of data and computer capacities, the methods could be applied more efficiently and successfully. The so-called deep learning approaches brought about an immense leap in quality. These massive gradual improvements to "machine learning on drugs" allow us to perceive a quasi-principle leap in AI that does not actually exist in this way. The systems are still learning according to certain rules and settings, patterns and distinctive features.

The next important step in the evolution of AI is the ability of the systems to learn autonomously and proactively to a wide extent. The first promising learn-to-learn approaches were applied in the AlphaGo example described. In addition, there are numerous promising research approaches in this area that will lead to algorithms adapting themselves or that will also develop new algorithms. This will, however, continue to happen in a rather formal-mechanistic understanding. This has little to do with a human's ability to learn. The next step of evolution, which then also contains human-like

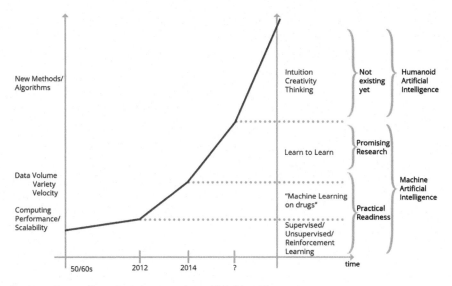

Fig. 2.4 Steps of evolution towards artificial intelligence

VISION ERROR RATE

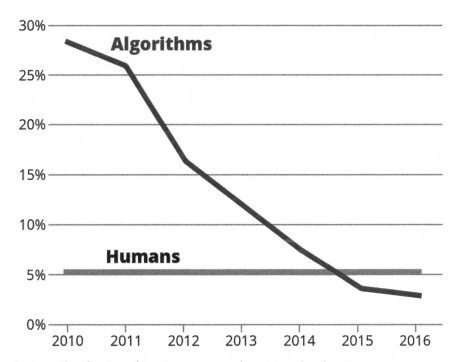

Fig. 2.5 Classification of images: AI systems have overtaken humans

abilities such as creativity, emotions and intuition, is a distant prospect and eludes a reliable temporal prognosis.

From a business point of view, this discussion may appear to be academic anyway. The decisive factor is the present-day perceived performance of the AI systems. And even today, they outperform human performance in many areas. Figure 2.5 shows the development of AI performance in image recognition. Even if the AI systems are still not perfect with their misclassification of 3% today, they have been outperforming the classification skills of humans since 2015. Thus, these systems can recognise the likes of reliable cancer diagnoses, fraud detection or other relevant patterns. This also applies to speech recognition.

Note

1. In contrast to conventional databases, data in this case is not kept on traditional hard drives but directly in the central memory. This significantly decreases the times of storing and accessing.

References

Duden.de. (2016). http://www.duden.de/rechtschreibung/Intelligenz.

Klug, A. (2016). *Assessment. Lexikon der Management-Diagnostik*. http://www.klug-md.de/Wissen/Lexikon.htm. Accessed 10 Jul 2017.

Manyika, J. et al. (2011). *Small States: Economic Review and Basic Statistics, Volume 17*. https://books.google.de/books?isbn=184929125X.

Rich, E., Knight, K., & Nair, S. B. (2009). *Artificial Intelligence* (3rd ed.). New York: Tata McGraw-Hill.

Russell, S. J., & Norvig, P. (2012/2016). *Artificial Intelligence—A Modern Approach*. London: Pearson Education.

Part II

AI Business: Framework and Maturity Model

3

AI Business: Framework and Maturity Model

3.1 Methods and Technologies

In the following, the various methods and technologies are briefly outlined and explained.

3.1.1 Symbolic AI

Since the conference at Dartmouth College in 1956, a variety of different methods and technologies have been developed for the construction of intelligent systems.

Even if neuronal networks and thus the approach of sub-symbolic AI dominates today, the field of research was dominated by the symbolic approach for a long time. This "classical" approach by John Haugeland called "Good Old-Fashioned Artificial Intelligence" (GOFAI) used defined rules to come to intelligent conclusions depending on the input. Up to the AI winter of the 1990s, "artificial intelligences" were developed by programming and filling control equipment and standards and databases to then be able to access them in practice. To this day, a large number of search, planning or optimisation algorithms and methods from the times of symbolic artificial intelligence are applied in modern systems, which today are simply regarded as excellent algorithms of informatics.

© The Author(s) 2019
P. Gentsch, *AI in Marketing, Sales and Service*,
https://doi.org/10.1007/978-3-319-89957-2_3

3.1.2 Natural Language Processing (NLP)

Computer linguistics covers the understanding, processing and generating of languages. "Natural language processing" describes the ability computers have to work with spoken or written text by extracting the meaning from the text or even generating text that is readable, stylistically natural and grammatically correct. With the help of NLP systems, computers are put in a position of not only reacting to formalised computer languages such as Java or C, but also to natural languages such as German or English.

A frequently used example of linguistics to illustrate the complexity of human language is the following: Every word in the sentence "time flies like an arrow" is distinct. But if we replace "time" with "fruit" and "arrow" with "banana", the sentence then says: "Fruit flies like a banana". Whereas "flies" in the first sentence still describes the verb "to fly", it becomes a noun in the second sentence "(fruit) flies" and the preposition "like"—"as" becomes the verb "to like" in the second sentence. Whilst a human intuitively recognises the correct meaning of the words, NLP uses a combination of different ML techniques to achieve the desired results.

Differences in performance become obvious in the own experiment of Google and Bing translation tools. Whereas the Google translator already works a lot and successfully with semantic ML methods, Bing still translates in many cases word for word.

Particularly topical in this field is the subject of speech recognition, which deals with the automatic transcription of human speech and, at present, is one of the major drivers of artificial intelligence in the retail market. At present, devices such as Amazon Echo, which are solely controlled by speech input, are already being sold.

A further application of computer linguistics lies in the field of "natural language generation" (NLG), e.g. in the automated writing of texts in strongly formalised areas such as sports or financial news. Other use cases are sentiment analyses in customer reviews, the automatic generation of keyword tags or the sifting through legal. The focus at present is on the use of chatbots in customer service and Conversational Commerce.

3.1.3 Rule-Based Expert Systems

Rule-based expert systems belong to one of the first profitable implementations of AI that are applied to this day. The fields of use are multifaceted and range from planning in logistics and air traffic over the production of consumer and capital goods down to medical diagnostics systems.

They are distinguished by the fact that the knowledge represented inside of them originates from experts (individual fields of expertise) in its nature and origin. Depending on the input variables, automatic conclusions are then derived from this knowledge. To this end, the knowledge (in the spirit of symbolic AI) must be codified, i.e. furnished with rules, and be linked to a derivation system to solve the challenges.

Frequently, the knowledge is derived from the factual database with the help of long chains of "IF-THEN rules". The advantage of expert systems lies in the fact that the formation of the results can be reproduced precisely by the user via the explanation components.

The ideas and knowledge about early knowledge- and rule-based expert systems are still applied today in modern systems. However, the knowledge must no longer be structured and stored in databases with great effort and in cooperation with experts, but can be captured and processed via natural language processing and machine learning methods in combination with great processing power in real time. Due to the sensation surrounding artificial neuronal networks, present-day systems are rarely advertised as expert systems However, they continue to be used frequently especially in medical applications.

3.1.4 Sub-symbolic AI

The approach of symbolic AI to systematically capture and codify knowledge was considered very promising for a long time. In a world that is being digitalised further and further, in which knowledge implicitly lies in the amounts of data, AI should be able to do something that knowledge-based expert systems inherently find difficult: Self-learning. Deep Blue, for example, was in fact able to beat Garry Kasparow in 1996 without the use of artificial neuronal networks, but only because the chess game had been formalised by humans and because the computer was able to compute up to 200 million moves per second from which the most promising one was then chosen.

In contrast to symbolic AI, the attempt is made in sub-symbolic AI to create structures with the help of artificial neuronal networks, which learn intelligent behaviour with biology-inspired information-processing mechanisms: It follows a bottom-up paradigm (Turing 1948). Many inspirations for mechanisms of this kind originate from psychology or even neurobiology research. This is why the term neural AI is sometimes used. The knowledge or the information is not explicitly readable, not like in the case of symbolic AI. With the help of the networks, the correlations to be studied are divided into sub-aspects and coded such that the mostly statistical learning mechanisms of machine learning can be applied (Russell and Norvig 2012). Sub-symbolic AI is thus an artificial neuronal framework for presenting problems for machine learning.

As can be seen in Fig. 3.1, every artificial neuronal network comprises an input layer (green), an output layer (yellow) and any number of hidden layers (blue), the number of which depends on the respective task. Each node, i.e. each neuron, within the system processes/adds the weighted input values from the environment or from preceding neurons and transfers the results to the next layer. An artificial neuronal network "learns" by the weighting of the connections of neurons to each other being adapted, new neurons being developed, deleted or derived from functions within neurons.

Even if artificial neuronal networks are nothing new, it has been possible in recent years to achieve great increases in performance due to the use of more efficient hardware and large amounts of data in combination with neuronal networks. In this context, the term "deep learning" is frequently mentioned, which describes the use of artificial neuronal networks with a wide number of hidden layers.

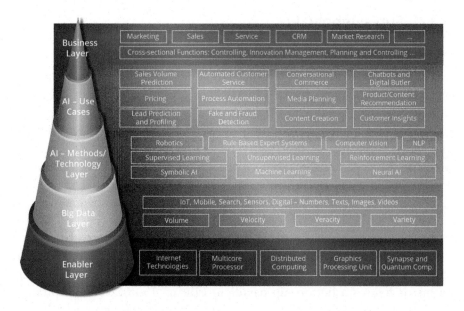

Fig. 3.1 Business AI framework (Gentsch)

At the end of 2011, a team of the Google X labs, of the US company's research department Alphabet, extracted around ten million stills from videos on YouTube and fed them into a system called "Google Brain" 16 with more than one million artificial neurons and more than a billion simulated connections. The result of the experiment was a classification of the images in various categories: Human faces, human bodies, (…) and cats. Whilst the result that the Internet is full of cats caused amusement, the publication also showed that with the help of particularly deep networks with a large number of hidden layers, technology is now capable of solving less precisely defined tasks. Deep learning enables computers to be taught tasks that humans intuitively find easy, such as the recognition of a cat, and which for a long time seemed to be only solvable with great effort in informatics.

3.1.5 Machine Learning

The term machine learning (ML) as a part of artificial intelligence is ubiquitous nowadays. The term is used for a wide number of various applications and methods that deal with the "generation of knowledge from experience".

The well-known US computer scientist Tom Mitchell defines machine learning as follows:

A computer program is said to learn from experience E with respect to some class of tasks T and performance measure P, if its performance at tasks in T, as measured by P, improves with experience E (Mitchell 1997).

An illustrative example of this would be a chess computer program that improves its performance (P) in playing chess (the task T) by experience (E), by playing as many games as possible (even against itself) and analysing them (Mitchell 1997).

Machine learning is not a fundamentally new approach for machines to generate "knowledge" from experience. Machine learning technology was used to filter out junk e-mails a long time ago. Whilst spam filters that tackled the problem with the help of knowledge modelling had to constantly be adapted manually, ML algorithms learn more with each e-mail and are able to autonomously adapt their performance accordingly.

Besides the fields of responsibility defined in the previous section, with machine learning, different ways of learning are differentiated from each other. The most common will be discussed in the following:

3.1.5.1 Supervised Learning

Supervised learning proceeds within clearly defined limits. Besides the actual data set, the right possible answers are already known. The supervised learning methods are meant to reveal the relationship between input and output data. These methods are used for tasks in the fields of classification as well as regression analyses. Regression is about predicting the results within a continuous output, which means that an attempt is made to allocate input variables to a continuous function. With the classification, in contrast, an attempt is made to predict results in a discreet output, i.e. allocate input variables to discreet categories.

The forecast of property prices, for example, based on the size of the houses, would be an example for a regression problem. If, instead of that, we forecast whether a house will cost more or less than a certain price depending on its size, that would be a categorisation, where the house would be placed in two discreet categories according to the price.

3.1.5.2 Unsupervised Learning

In contrast to supervised learning, with unsupervised learning, the system is not given target values labelled in advance. It is meant to autonomously identify commonalities in the data sets and then form clusters or compromise the data. As a rule, it is about discovering patterns in data that humans are unaware of.

Unsupervised learning algorithms can, for example, be used for customer or market segmentation or for clustering genes in genetic research, to reduce the number of characteristic values. With the help of this compression, computing could be faster afterwards without loss of data.

3.1.5.3 Reinforcement Learning

An alternative to unsupervised learning is provided by the models of reinforcement learning where learning patterns from nature are reproduced in concepts. Through the combination of dynamic programming and supervised learning, problems that previously seemed to be unsolvable can be solved. Differently to unsupervised learning, the system does not have an ideal approach at the beginning of the learning phase. This has to be found step by step by trial and error. Good approaches are rewarded and steps tending to be bad are sanctioned with penalisation. The system is able to incorporate a multitude of environmental influences into the decisions made and to respond to them. Reinforcement learning belongs to the field of exploration learning, where a system autonomously, thus irrespective of the rewards and penalties pointing in the right direction, has to find its own solutions that can be clearly differentiated from those thought up by humans. Reinforcement learning attracted a notable amount of attention after the victory of Google DeepMind's AlphaGo over Lee Sedol. The system used applied deep reinforcement learning among others to improve its strategy in simulated games against itself. Through reinforcement learning, artificial intelligences thus acquire the ability to find new approaches on their own and to at least seemingly act intuitively.

3.1.6 Computer Vision and Machine Vision

Computer vision describes the ability of computers or subsystems to identify objects, scenes and activities in images. To this end, technologies are used with the help of which the complex image analysis tasks are divided among as small sub-tasks as possible and then computed. These techniques are applied to recognise individual edges, lines and textures of objects in one. Classification, machine learning and other processes, for example, are used to determine whether the features identified in an image probably represent an object already known to the system.

Computer vision has multifaceted applications, among them the analysis of medical imaging to improve prognoses, diagnoses and treatment of diseases or facial recognition on Facebook, which ensures that users are automatically recognised by algorithms and are suggested for tags. Such systems are already used for security and surveillance purposes for the identification of suspects. In addition, e-commerce companies such as Amazon are working on systems with which specific products can be identified on images and subsequently be purchased directly online. Whilst researchers in the field of computer vision are working on the aim of being able to utilise systems independent of the environment, with machine vision, sensors are used with the help of which relevant information can be captured within restricted environments. This discipline is technically mature to the extent that it is no longer part of ongoing informatics research, but part of system technology today. At the same time, it is less a matter of recognising the meaning or content of an image but of deriving information relevant for action.

3.1.7 Robotics

The interdisciplinary interplay of mechanical and electrical engineers with information scientists is what makes robotics possible in the first place. The combination of various technologies such as machine learning, computer vision, rule-based systems as well as small, high-performance sensors has led in recent years to a new generation of robots. In contrast to the famous industrial robots of the automobile industry, which are utilised for simple mechanical tasks, more recent models can work together with humans and adapt flexibly to various tasks.

3.2   Framework and Maturity Model

In this chapter, the bridge to business is built via the use cases. The subjects of framework and maturity model will be discussed. It will be explained how the set-up of a framework depends on the relationship of the individual areas with each other. Big data and AI layers are thus first made possible by the enabler layers. The AI use cases, in contrast, have a direct influence on the business layer. The layer model presented accommodates these dependencies. In addition, the various phases on the way to an algorithmic enterprise are presented as degrees of maturity. The model shows the different steps of development from the non-algorithmic enterprise over the semi-automated to the automated enterprise. The super intelligence enterprise represents the highest degree of maturity. Finally, the benefits and purpose of a maturity-level model are discussed.

In the last part, the question is answered as to who is in charge of the establishment of AI and the transformation to an algorithmic business.

3.3   AI Framework—The 360° Perspective

3.3.1 Motivation and Benefit

After the presentation and explanation of the enabler technologies and AI methods (Sect. 3.1) in this chapter, the bridge to business is to be built via the use cases. The way the set-up of the framework depends on the relationship the individual areas have with each other will be explained. Big data and AI layers are thus first made possible by the enabler layers. The AI use cases, in contrast, have a direct influence on the business layer. The layer model presented accommodates these dependencies.

Within the AI business framework, the relevant topics and terms are systemised, categorised and linked up to each other. The AI framework thus acts as a transmission belt of the factors of success and drivers of AI in companies down to the operational applications.

The AI business framework demonstrates the entire range of tools and solutions and is thus meant to enable a better orientation in the jungle of artificial intelligence. An ever unambiguous assignment of data, technologies, methods, use cases and operational applications is not possible. The correlations are far too complex and multifaceted.

3.3.2 The Layers of the AI Framework

The factors of success of AI were already described in the previous chapters. In the framework, these are presented in the bottom layer, the so-called enabler layer. Due to their contribution towards the development of AI and the emergence of big data, the Internet technologies, multi-core processors, distributed computing, GPUs as well as the future technologies and synapsis and quantum chips were adopted in the framework. The significance of big data for the current development of artificial intelligence is accommodated with its own level (Fig. 3.1).

Particular attention within this layer is given to the following:

- Structured and unstructured data (variety). As already described in Sect. 2.1, the methods originating from AI research that go beyond the analysis of structured data also enable the machine-processing of unstructured data.
- Large amounts of data for the training of machine learning algorithms (volume) are decisive for the development of AI.
- The speed (velocity) in combination with the amounts of data with which data is generated and evaluated can no longer be mastered by human actors without the help of intelligent systems. ML algorithms help to master the flow of data and to separate the important from the unimportant.
- It is now very difficult to determine the credibility of the data (veracity) manually. At present, systems that are meant to distinguish between real news and fake news are being worked on.
- Data sources are still being shown as their own item in the framework: Whether from the Internet of Things (IoT), mobile end devices, search applications or other digital applications. Data is the fuel for the AI machine. Of significance is neither its origin nor its structure, nor can there be "too much" data nowadays.

3.3.3 AI Use Cases

For the layer in which the business and AI world are united, the "artificial intelligence use cases layer", a multitude of current and future examples can be found.

The use cases as a further layer for the AI business framework are presented and explained in the following (Fig. 3.2).

3.3.4 Automated Customer Service

In correlation with the developments of the personal assistants, the customer service departments of companies can be organised significantly more efficiently thanks to the advances in computer linguistics. Whereas the customer experience nowadays frequently turns out to be negative during calls with answers like "sorry, your question was not understood, did you mean…?", NLP algorithms help such experiences to be a part of the past and simple issues can actually be explained easily in natural languages (cf. Sect. 8.2).

3.3.5 Content Creation

Content marketing and the relevant addressing of target groups have long been preached as the formula for success in marketing. However, as a rule, the potential of digitally available data for the automatic creation of content is not made use of. Algorithms can, for example, gain interesting and unfalsified insights on the basis of public Internet data in real time. The likes of automatic infographs can be generated, for example, which demonstrate business development depending on the application of certain technologies, on the digital maturity or on the use of advertising drive. Equally, new

Fig. 3.2 Use cases for the AI business framework (Gentsch)

market developments and upcoming topics can be automatically recognised on the basis of big data. Topical discussions and reports can thus be used systematically and quickly ("news-jacking"). The editorial description and explanation of the insights generated is covered by a suitable analysis team. This is where computer linguistics, to be more specific natural language generation, is applied. What is meant by this is systems that create texts based on figures and individual facts. It is difficult to differentiate these from texts written by a human. Due to their consistent structure, they are particularly suitable for sports or financial news.

3.3.6 Conversational Commerce, Chatbots and Personal Assistants

Instead of artificial interfaces such as websites and apps, customers can communicate with company systems via totally natural communication as in spoken or written language. This is facilitated by the developments in computer linguistics previously described. This type of communication also enables less technology-affine people to deal with new technologies—at present, various providers are competing for the best personal assistants. And for good reason, too: Companies that assert themselves here and which are able to sell their solution to customers will develop a kind of portal for other companies in the medium term to sell their products to customers. This is why this topic is near the top of companies' agendas, companies such as Amazon, Apple or Google (cf. Sect. 4.5 "Conversational Commerce and AI in the GAFA Platform Economy").

3.3.7 Customer Insights

One of the key tasks of classic market research is the systematic deduction and explanation of "how customers work"—called customer insight. In order to obtain feedback from customers on products, classical market research avails of extensive tools: Focus groups, customer surveys, panels, etc. The main disadvantage of this primary research is the effort involved. On the Internet, for example, thousands of product reviews can be analysed automatically at any time: Ratings and reviews that are distributed across various Internet platforms are captured and integrated intelligently by bots. With the help of NLP, the key customer statements are automatically retrieved from the free texts of the reviews. In order to gain more in-depth insights, the insights gained have to be correlated with other data such as complaints, sales or customer satisfaction.

3.3.8 Fake and Fraud Detection

AI has been used for some time now in detecting and predicting fraud. In the area of marketing and communication, fake news and manipulation by way of targeted disinformation is under discussion. The use of (chat)bots for targeted promotion, disinformation and manipulation harbours a high risk for companies. But the topic is nothing new. In the past, many companies have contracted agencies to remove or paper over negative posts in social networks to push topics or submit positive feedback or negative for the competitor. Some companies did not survive when exposed or suffered a damaged reputation or had to endure bad shitstorms. We have equally been occupying ourselves for a long time with the phenomenon of astroturfing. This process cannot be automated and scaled. And even here, algorithms and AI can help. A systematic, data-driven approach can automatically recognise patterns from manipulative bots, e.g. posting frequency and times, network of followers, contents and tonalities. Modern AI methods are being used here for detection and prevention. These methods are already being successfully used for click and credit card fraud. In a way, we are beating the manipulative bots with the same weapons they are using for automatic disinformation and manipulation.

3.3.9 Lead Prediction and Profiling

AI enables the automatic recognition and profiling of potential customers. For example, new customers and markets can be identified and characterised on the basis of given customer profiles via so-called statistical twins. In doing so, the selected companies were packed with thousands of attributes for a digital signature. On the basis of these data vectors, new customers can be predicted in digital space using AI algorithms (predictive analytics). Leads and markets that do not match the classical acquisition strategy, but represent potential buyers can also be identified with it—communication potentials beyond the antiquated industry and segment perspective (cf. Sect. 5.1 "Sales and Marketing Reloaded—Deep Learning Facilitates New Ways of Winning Customers and Markets").

In addition, communication and sales triggers can be identified and evaluated by way of dynamic profiling: With which event is the sales approach particularly successful? Time- and context-specific sales signals significantly increase the probability of conversion. Moreover, the trigger can be used as a reason for communication for the right sales pitch. Besides the addresses of the companies, leads for the right means of communication can also be supplied at the same time. In some cases, a direct approach on Xing and LinkedIn is more promising than a phone call or an e-mail.

3.3.10 Media Planning

The media market has been distinguished by self-serving and interest-driven plans and argumentations for years now. Algorithm-based technology platforms enable transparent and efficient media planning on the basis of artificial intelligence. AI and algorithms can capture a multitude of relevant active and reactive media data points and automatically assess them subjectively. This way, the so frequently subjective and self-interest-driven planning experiences an empirical earthing and validation (cf. Sect. 5.8 "The Future of Media Planning").

3.3.11 Pricing

The use of AI software to determine retail prices for all goods from fuel over office supplies down to food is growing progressively. At the same time, it is not about how competition changes the prices.

The AI algorithms analyse thousands of data points on a continuous basis and calculate prices the software believes the consumers are willing to pay. In other words: It is all about the search for the ideal price—not the lowest one. AI pricing software analyses huge amount of historical and real-time data and attempts to establish how consumers will react to price changes under certain scenarios. Tactics are updated on the basis of experience. These solutions equally try to learn and take into consideration human behaviour.

All in all, it is not about getting more money out of the customers. It is rather about making margins with customers who do not care about them and to go without margins with customers who do.

The use of AI pricing algorithms is constantly growing in Europe and the USA, at petrol stations in particular. The approach is interesting, especially for retailers. Staples, for example, uses AI to post the prices of more than 30,000 products on their websites every day.

The mother of all online retailers in the USA, Amazon and their third-party providers, were among the first to utilise dynamic pricing, a precursor of AI pricing. Today, Amazon uses AI technologies to a great extent to skim the maximum consumer surplus (also see the practical example in Sect. 5.10 "Next Best Action—Recommender Systems Next Level").

3.3.12 Process Automation

The subject of process automation is nothing new. It was discussed intensively and implemented in the 1990s in the scope of the so-called business process management/reengineering. The focus of this was more on industrial and production processes and less on marketing and sales processes. In addition, the algorithmic support was mostly classically rule-based.

Robotic Process Automation (RPA) is a software automation tool that automates tasks such as data extraction and preparation as a matter of routine. The robot has a kind of user ID and can perform rule-based tasks such as accessing e-mails and other systems, make calculations, create documents and reports and revise files. RPA has, for example, helped a large insurance company to reduce hold procedures that affected 2500 high-risk accounts per day. This meant that the pressure was taken off 81 percentage of the staff, who were then able to focus decidedly on proactive account management (Mckinsey 2017).

Thanks to modern AI algorithmics, significantly improved efficiency, increased staff performance, a decrease in operational risks and an optimised customer experience could be achieved due to the intelligent process automation.

3.3.13 Product/Content Recommendation

Frequently, recommendations for products and/or content are proposed and managed manually by editors and shop managers. This is, however, very time-consuming and badly scaled. A modern web shop cannot be imagined without recommendation engines for personalised recommendations today. Whilst simple algorithms of the shopping trolley analysis were used in the early days—"Customers who bought product A also bought product B"—, today, AI methods, which taken into consideration a multitude of data points, are increasingly applied.

On the basis of their clicking and purchasing behaviour, the user is, for example, shown additional matching content to better satisfy their interest and to create additional buying incentives. A particularly promising approach is based on AI reinforcement learning.

The open source approach of the GAFA world is also of interest here. Similar to how Google went public with the AI deep learning framework Tensorflow, Amazon 2016 launched at the same time DSSTNE (pronounced "destiny") as an open source framework that companies can use for their product recommendations. It may seem astonishing at first that the "inventor" of the automatic product recommendation makes their core asset available to the community, yet the rationale is clear after all as per Amazon's corresponding FAQs: "We hope researchers from all over the world will work together to be able to further improve the recommendation system. But more importantly, we hope that it triggers innovations in many other areas".

3.3.14 Sales Volume Prediction

The sales volume prediction is decisive for most companies, but it is also a difficult field of management. Most researchers and companies use statistical methods such as the regression analysis to forecast and analyse sales volumes.

Moreover, as a rule, only very small amounts of data are used for forecasts for sales. To increase the quality of the sales forecast, numerous other data points can be considered by AI. These include both historical and real-time data, internal and external data, economic and environmental data, micro-economic and macro-economic data points (sales figures, warehouse data, prices, weather, public holiday constellations, competitor prices, etc.). Algorithms and AI, on the one hand, assist in capturing these many structured and unstructured data points systematically automated and, on the other hand, to automatically analyse them for an accurate forecast. One of the most well-known systems in Germany is the Blue Yonder solution.

3.4 AI Maturity Model: Process Model with Roadmap

3.4.1 Degrees of Maturity and Phases

In Fig. 3.3, the various phases on the path towards the algorithmic enterprise are presented as the degree of maturity. The model shows the various stages of development from the non-algorithmic enterprise via the semi-automated to the automated enterprise. The super intelligence enterprise represents the highest level of maturity. This is where the autonomous and self-learning AI systems described in Sect. 3.1 are used. This highest degree of maturity is difficult to forecast due to the uncertainty of the time of occurrence of the singularity and it is not of relevance in the short or medium term. According to the various expert opinions, this highest degree of maturity of AI is to be expected between 2040 and 2090. The individual degrees of maturity are described in the following:

Data, algorithms and AI do not play a business-critical role when it comes to the non-algorithmic enterprise (Fig. 3.4). The topics are ascribed rather an operative and transactional significance. The strategy and organisation are rather classical and less analytical and data-driven. Upon the transition to a semi-automated enterprise, the crucial value of algorithmics and AI is increasingly recognised. Accordingly, there are corresponding data and analytics structures. Characteristic is the increased degree of automation of data collection and analysis as well as the decision-making and implementation (Fig. 3.5).

This is made possible by a holistic integration of data sources, analyses and process chains. Data, analytics and AI facilitate the creation and implementation of new business processes and models in this maturity level. The data- and analytics-driven real-time company obtains systematic competitive advantages this way (Fig. 3.6).

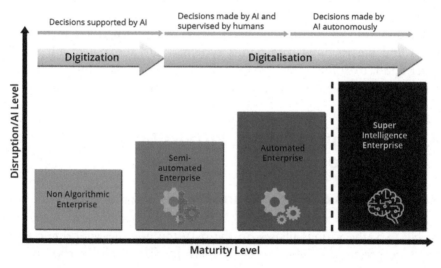

Fig. 3.3 Algorithmic maturity model (Gentsch)

Whilst with the automated enterprise the approaches of narrow AI described in Chapter 2 are applied, the super intelligence enterprise concludes the potential of autonomy and self-learning of companies by way of general and super intelligence. This scenario currently appearing to be hardly realistic has two types of manifestation. In the positive version, we as humans control the framework conditions and rules of the autonomous AI systems. We can intervene and rectify via regulative and corrective measures at any time. Productivity and well-being are increased further by the performance, scalability and innovations of these super intelligences. In the negative version, we as humans have lost control over the framework conditions and rules of the autonomous systems. There is no longer the last

Non-Algorithmic Enterprise

Strategy

- No decided AI/algorithmic strategy
- Data is not regarded as critical to success
- No alignment with goals (marketing, sales, service)
- Analytics is part of the IT

People/Orga:

- No CDO
- No Data Scientists
- Limited analytics talents
- Classical marketeers

Decisions:

- Decisions made solely by humans
- No automation of processes
- Only rule-based systems are applied (e.g. controlled paths in service)
- Hands-on mentality

Data:

- Data is used in operative systems/transaction-oriented
- Focus is on structured data
- The different data sources are not linked up with each other
- Data is not captured and used systematically
- No automation of data collection and analysis

Analytics:

- Simple analytics (XLS, SPSS, ...)
- Isolated analytics (web analytics, offline analytics, ...)
- Ad-hoc analytics/ex-post analytics: What happened?

Fig. 3.4 Non-algorithmic enterprise (Gentsch)

Semi-Automated Enterprise

Strategy

- Rudimentary AI/algorithmic-strategy
- Data is regarded as business-relevant
- Alignment with partial goals (marketing, sales, service)
- Analytics is part of the IT and specialist departments

People/Orga:

- Typically no CDO
- Analytically oriented staff
- Cooperation between IT and marketing, sales, service

Decisions:

- Algorithms recommend course of action
- Humans make final decisions
- Algorithms make partial decisions; humans have to confirm them before they are executed
- Automation of individual processes (marketing automation, service and chatbots …)
- Both rule-based and knowledge-based systems are applied (e.g. Knowledge database for call centre agents)
- Product recommendation systems
- Marketing automation/drip campaigns

Data:

- Data is also used strategically (customer value-related segmentation, sales prediction)
- Focus is on structured and unstructured data
- Data source are partially linked up with each other
- Capture of various touchpoints of the customer journey
- Data is partially captured and used systematically
- Partial optimisation of data collection and analysis

Analytics:

- Advanced analytics; data mining/machine learning: "Why did something happen?" or "What is expected to happen?"
- A/B testing
- Partially integrated analytics (analysis via various touchpoints of the customer journey)
- Ad-hoc/ex-post-analytics: What happened?
- Analysis models are not integrated automatically in processes

Fig. 3.5 Semi-automated enterprise (Gentsch)

call for man. AI systems further develop uncontrolled without the possibility of human intervention—permanently and with an open-ended result (Fig. 3.7).

Even if the super intelligence enterprise seems to be a long way away, there are some businesses today with an extremely high level of automation.

Automated Enterprise

Strategy

- Decided AI/algorithmic strategy
- Data and analytics for business processes and business models
- Data as value drivers and competitive advantage
- Predictive analytics are used to optimise and automate company decisions
- Alignment with targets (marketing, sales, service)
- Analytics is primarily part of the specialist departments/IT as supporter

People/Orga:

- CDO
- Data-driven Mindset
- Marketing, sales, service are in the data/analytics driving seat
- Chief Conversational Officer responsible for automation/speech control of customer interaction

Decisions:

- Routine decisions are made and executed by algorithms
- The majority of decisions are made by algorithms; partially directly executed; partially confirmed by humans prior to execution
- Humans have the last say in
- Automation of the entire customer journey (automated control of the entire customer journey ...)
- Besides rule-based and knowledge-based systems, AI systems are also used
- Integrated recommendation system for products, content, communication, ...
- Integrated decision chains (CRM, customer journey, marketing automation, ...) automatic identification, profiling and addressing target groups
- Automation of the entire sales funnel
- Automation of content generation/curation
- Internal planning and coordination processes are automatically controlled by bots (conversational office)

Data:

- Data is also used strategically (customer value-related segmentation, sales prediction
- Focus is on structured and unstructured data
- Relevant data sources are fully linked up with each other
- Collection of all relevant touchpoints of the customer journey
- Data is automatically collected and used

Analytics:

- Advanced analytics; data mining/machine learning /deep learning/narrow AI: "What should be done"
- Analytics results are automatically used for creating and optimising business processes
- Automated attribution modelling optimises the customer journey in real-time
- Integrated analytics (analysis via various touchpoints of the customer journey)
- Ad-hoc/ex-post analytics: What happened?
- Analysis models are not integrated automatically in processes and executed
- Automated analysis/creation of content

Fig. 3.6 Automated enterprise (Gentsch)

Guests at the Henn-na Hotel (http://www.h-n-h.jp/en) in Japan, for example, are greeted by a multilingual robot who helps the guests to check in and out. Artificial butlers take the luggage to the rooms and there is a room for the storage of luggage which is put away by a mechanical arm. The devices are not gimmicks for the company but a serious effort to be more efficient. The hotel is keyless and uses facial recognition technology instead of normal electronic key cards. A guest's photograph is taken digitally at check-in. In the rooms themselves, a computer globe with a stylised face caters for the comfort of the guests. The computer globe can be used on the basis of digital butler technology (Sect. 4.1) to switch the light on and off, to enquire about the weather or a suitable restaurant.

Amazon can be quoted as a company with a high maturity model. It has a high level of maturity across all dimensions (Fig. 3.8).

DAO (decentralised autonomous organisation) is a highly automated and virtual organisational construct. This is a virtual company without a business domicile, CEO or staff, which organises itself with the help of codes.

Super Intelligence Enterprise

AI systems not only execute algorithms but rather AI develops its own algorithms autonomously and flexibly. Strategy, resources, data and analytics are the subject of planning and execution processes of autonomous systems.

Positive Scenario	Negative scenario
Humans control the framework conditions and control the autonomous AI systems. It is always possible to intervene via regulative and corrective measures. AI systems increase productivity and well-being through performance, scalability and innovations.	Humans have lost control of the framework conditions and rules of autonomous AI systems. AI systems are developing uncontrolled without the possibility o human intervention. The last can of humans and final red button have become obsolete. It is questionable whether the developments resulting from this are for the well-being of man.

Fig. 3.7 Super intelligence enterprise (Gentsch)

DAO broke all crowdfunding records in as early as 2016 and collected 160 million US$. DAO works like an investment fund, whereby the collected capital is invested into start-ups and products to yield a profit for the members of the organisation. The so-called crowdfunders vote on the direction in which the organisation is to develop.

So-called smart contracts regulate the investments of the DAO members. These are algorithms added to the software, which automatically and permanently review the terms of a contract and take corresponding measures. These rules are stored in a decentral managed database—the so-called blockchain.

When the defined goal has been achieved, the smart contract automatically executes the transfer. The DAO members receive tokens for the voting, which are used for voting, in line with the money paid in. In addition, the members can also submit their own ideas for projects and ideas to be financed by the DAO.

DAO automates company processes on the basis of blockchain technologies. The governance rules are executed by the "algorithmic CEO" and not, as is customary, by the Board of Directors. A company organisation is formed that is fully digitalised.

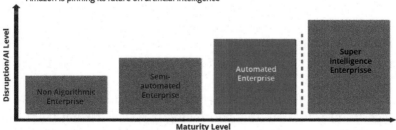

Fig. 3.8 Maturity model for Amazon (Gentsch)

If we follow the definition of contract theory, according to which a company is nothing other than a network of contracts in which objectives, authorisations and terms are laid down, the high level of automation of company processes and decisions seems realistic. Employment contracts, for example, regulate and control the actions of the employees. Employees "execute" tasks laid down in the contract. The title CEO—Chief Executive Officer—is derived from this execution rationale. Contracts thus regulate everything in a company, why not be executed by algorithms instead of humans?

Algorithmic technology has the potential to fundamentally change the way we do business, and has been flagged as the most prominent sweeping change since the industrial revolution (Charmaine Glavas, Queensland University of Technology, 2016).

3.4.2 Benefit and Purpose

The concept of a maturity-level model not only has the aim of classifying companies into individual levels but moreover indicates a road that companies have to take in competition. Before companies occupy themselves with AI, they should digitalise and structure their processes systematically. Benefit and purpose can in principle be subdivided into three types:

Descriptive is a maturity-level model to the extent that a descriptive classification takes place. This helps to obtain a better understanding of the current situation. This allows companies, for example, to recognise the status quo regarding a certain topic.

In addition, a maturity-level model provides the possibility of a normative character. The recognition of the current state is obtained by the constructive maturity levels of the model. The maturity-level model is ground-breaking if it indicates what is necessary to achieve future or higher degrees of maturity.

A further benefit of a maturity-level model is that it can be applied in a comparative way. The position or the maturity level within a model can be compared. This facilitates the execution of an internal and external analysis. On the one hand, this facilitates the comparison of company-internal departments; on the other hand, the company can be measured with the competitors in competition.

All in all, companies can locate their current status with regard to big data, algorithmics and AI. This positioning is a vital starting point on the systematic path to becoming an algorithmic business. On the basis of the positioning, targeted measures can be derived for the next highest maturity level. Furthermore, benchmarking helps in and beyond sectors (Fig. 3.9).

3.5 Algorithmic Business—On the Way Towards Self-Driven Companies

The effects and implications of algorithmics and AI affect the entire corporate value added chain. According to the focus of the book, the "business layer" of the AI business framework has foregrounded the "customer facing" processes and functions. In this chapter, the potentials for the entire corporate value creation are briefly described. It will be shown that artificial intelligence can change the way of working in classical company areas both sustainably and radically: By using artificial intelligence, companies can not only exploit efficiency and productivity potentials but also cater better to customers and thus create added value. In addition, the significance of the ideas and potentials of so-called Conversational Commerce (Sect. 4.2) for internal company functions and processes will be illustrated and explained (Conversational Office). Finally, the areas of marketing, market research and controlling (as relevant cross-sectional function) will be described and explained in more detail. Furthermore, algorithmics and AI also have the

 Objective positioning and assessment of past measures in the field of big data, algorithmics and artificial intelligence.

 The identification of company-specific strengths/weaknesses and the direct comparison with relevant competitor companies and industrial benchmarks.

 Questioning of established processes, though structures and measures in the field of big data, algorithmics and artificial intelligence.

 The identification of necessary measures, prioritising the measures and deriving a roadmap in which both development initiatives and quick wins are identified.

 On the basis of the roadmap, technology, financial and staff resources can be allocated ideally according to the priorities, necessary requirements can be created (enablers) and concrete measures can be developed that pay directly into the most important company KPIs (sales, profit, reach, relevance, etc.).

 Reference model for the sustainable control of progress and control of the digital transformation to an algorithmic business with time.

Fig. 3.9 The benefit of the algorithmic business maturity model (Gentsch)

Fig. 3.10 The business layer for the AI business framework (Gentsch)

potential of reinventing business models; these topics will also be treated in this chapter. Finally, it will be investigated whether it makes sense to install the position of a chief artificial intelligence officer in companies.

3.5.1 Classical Company Areas

The fact that artificial intelligence will change the way of working sustainably and radically can be demonstrated in the following fields of application. By using artificial intelligence, companies can not only exploit efficiency and productivity potentials but also, as described above, cater better to customers and thus create added value. This issue is frequently underestimated in the discussion about AI in the corporate world. Employees in companies will have to learn to work together with smart technologies. Whilst well-structured and standardised areas of artificial intelligence can be adopted, there will be a continued necessity for human staff in areas where empathy or the collaboration with humans is involved. There is thus more than only competitive advantages when reducing staff and increasing productivity. Further, it is not necessarily a given that the use of AI is more efficient than a conventional employee. The development of artificial intelligence has indeed become more affordable than a few years ago due to open source frameworks, yet statements on the economic feasibility of AI cannot be made across the board (Fig. 3.10).

3.5.2 Inbound Logistics

Inbound logistics are the first primary activity of a company's value added chain. The most important tasks of logistics include accepting goods, controlling stocks and warehousing. Companies are working on optimising the processes in their warehouses with the help of intelligent software. Examples for the use of artificial intelligence are shown in the logistics centres of the Japanese electronics groups Hitachi or Zappos. Even the online retailer Amazon uses AI technology, starting with the takeover of "AIva Robotics" in 2012. AIva endeavoured to create better logistics solutions for online retailers. On this basis, today's "Amazon Robotics" strives to produce robots that contribute towards automatic process flows in the logistics centres. In 2014, Amazon introduced "Alva Robotics" for the first time in California, as a trial run at first. In the meantime, the robots are being used as standard in the USA as well as in Europe. The robots move at a speed of about 5.5 km/h and weigh approx. 145 kg. They can lift up to 340 kg in weight. Together with the intelligent software, the robots are to form an automated logistics process. The scenario looks like this:

At the point of acceptance, the goods are accepted from the delivery man. There, the software gives each product a code for it to be found again. After that, the goods are placed "chaotically" on the warehouse shelves—wherever there happens to be a space for them. The aim of this is to be able to find articles at several places in the warehouse to keep walking distances as short as possible. The ordering and warehouse management system knows exactly where the individual articles are and what the best way is to transport them. As soon as the computer system receives an order, the electronically equipped commissioner moves to the shelf where the products are located and lifts them up to then take them to the desired packing station. In the process, the system informs of the nearest place on the shelf and the shortest distance to the station. At the packing station, the shelves are put down so that the staff can take the products needed and pack them.

The product code contains important product-specific data that is captured by the scan in the system. The intelligent software that analyses orders in real time and takes care of all processes finds the product again on the basis of this. With the help of intelligent algorithms, the management system not only calculates the shortest distance but also makes sure crashing is avoided. With intelligent robot and warehousing systems, Amazon would like to effectively catch up on the increase in orders. The aim is to not only render services to the customers speedily and reliably, but to also secure effective and easy work for the staff. According to Roy Perticucci, Amazon's Vice President Operations in Europe, roots taking over warehousing tasks leads to more products being delivered in shorter times. The reason for this is the shorter distances which, in turn, lead to shorter delivery times.

In some cases, orders that used to take hours to process can now be processed within minutes. Moreover, the accident rate in the warehouse is decreasing to a constantly low rate. Furthermore, it should be possible to store 50% more goods, at the same time, the costs in the warehouses are said to have decreased by 40%.

With the increase in the robot-controlled logistics chain, the constant increase in efficiency is also to be expected. The online retailer pursues the desire to fully automate the logistics chain. In addition to Amazon, the electronics group Hitachi also relies on AI software. The program analyses the way the staff work in detail and compares this with new approaches. At the same time, the software establishes how a work process can be integrated most effectively and gives the staff instructions. The group states that the AI system continuously analyses data and constantly learns something new about the warehouse processes. In addition, Hitachi stated that warehouses with artificial intelligence exhibited an 8 percentage increase in productivity in comparison with normal locations. Even if the program gives instructions by way of the big data analysis, it could equally integrate new approaches by way of optimised processes. After use in logistics, Hitachi hopes that AI will improve additional work processes in other areas.

How human employees find such a standardisation is debatable. Monitoring and controlling leads to a restriction in the staff's freedom which can cause mental problems and demotivation, Jürgen Pfitzmann, work organisation expert at the University of Kassel believes. Dave Clark, Amazon's head of global logistics defends the way of working according to strict instructions. In the same way as many companies, Amazon also has strict expectations of their staff. They seek to adapt target figures to local circumstances to not ask too much of individuals. De-facto work is long-term and predictable. A flexible and efficient process is targeted, which contributes towards the ability to respond more quickly to social change. All in all, robots and AI-shaped systems improve the logistics processes and facilitate fast responses to certain problems. If we consider that in the past fewer potentials for optimisation were possible in logistics processes, advancing technology today provides new opportunities for companies. Amazon is a leading example of innovations. The online retailer has been hosting the Amazon PicAIng Challenge since 2015. With this competition, teams from universities and companies can compete against each other with robots they have built themselves. "The aim of the advertised 'Amazon Robotics' PicAIng Challenge is to intensify the exchange of know-how for robotics between science and business and to promote innovations of robotics applications within logistics". Yet, although Amazon would like to utilise more robots, humans are still of great significance for the enterprise, as robots need the experiences of the staff to acquire knowledge that they can use, especially as the systems are also monitored and partially controlled.

3.5.3 Production

In classical industrial production such as in the car-manufacturing industry, the effects of AI and robotics can already be felt. The previously very structured processes can be digitalised and automated comparably fast. As a result, not only increases in productivity but also improved control options as well as constantly high quality can be achieved.

Terms such as "smart factory" stand for the machine's own decisions as to what they want to manufacture and when, and for much more. Indeed, some steps still need to be initiated for the vision of automated and intelligent production, yet research organisations have long been working on solutions for partial areas to alleviate the way humans work and improve processes.

3.5.4 Controlling

Companies can also be monitored and controlled more efficiently by using algorithms, as some of the tasks to be executed manually can be taken over by AI systems. Even the quality and speed of controlling can be increased by using intelligent algorithms.

3.5.5 Fulfilment

Nowadays, the entire value added chain from accepting an order over warehousing and commissioning down to dispatch is frequently contracted to specialised fulfilment service providers. Industry giants like Amazon or DHL have been working consistently for years on the improvement in their processes and, in the meantime, are employing robots in warehouses, for example, to increase efficiency or they have the latest algorithms plan their tours. Even if these processes already are highly developed, they still cannot be implemented to this day without human intervention.

3.5.6 Management

Whilst the creation and analysis of reports or target and resource management can be strongly supported or even completely taken over by machines, tasks such as drawing up strategies or leading employees are still carried out in the long term by managers. The challenges for the business management and administration will be to utilise the accomplishments of AI in such a way that as high an added value as possible is generated for the company.

3.5.7 Sales/CRM and Marketing

In these fields considerably more can be achieved by the application of artificial intelligence than just increasing efficiency. Personalised, custom-made product and price combinations for every customer can be implemented with the help of artificial intelligence. Thanks to modern algorithmics, personalised advertisements in online marketing are standard nowadays.

3.5.8 Outbound Logistics

The most significant task of outbound logistics is the distribution of the products. Artificial intelligence opens up new opportunities in logistics and is posing new challenges to the companies. The transformation demands dynamic and self-controlling processes that are based on intelligent consignments. The potential for the use of learning machines in logistics is significantly high. AI is not only meant to cooperate with humans without problems, but also recognise routine tasks and be able to learn them by drawing its own conclusions. Example Amazon: Here, this data is based on customer experiences and evaluations by staff in the logistics centres. The software in the packing area, for example, from the interface for all incoming information regarding the product. Data flows from various sources into the system. This includes customer reviews that relate to the packaging in particular. Customers can, for example, submit a review on the service and product quality as well as on the packaging.

Criticism concerning the unsuitable size or inadequately packed goods is analysed by the system and evaluated. Furthermore, the software filters field reports by the staff that are based on insights from daily routine. The system also captures important key data relating to the height, length and width and weight. The software recognises patterns in the data and selects the right size of packaging on this basis.

The Asia-Pacific Innovation Center of DHL in Singapore is occupying itself with innovative logistics solutions by way of artificial intelligence and robot technologies. At the centre, one can watch "Mr Baxter" at work. Mr Baxter collects the parcels from the warehouse shelf and stacks them onto a vehicle. The sensor-controlled vehicle transports the consignment to another part of the warehouse. Baxter enables another human-robot interaction—he stops the minute somebody approaches him. In practice, the robot is currently being tested at DHL along with another robot, "Sawyer". Due to the further development towards collaborative robots, the area of application has been extended. Besides the job of moving parcels elsewhere, the two perform packing tasks or labelling for shop sales. The high-performance and intelligent robots take on tasks that used to be difficult to automate.

In the meantime, artificial intelligence is also being used for the carriage of goods because not only the constantly increasing number of orders and parcels is a challenge for companies, but also the increasing competing for customers. Online retailers in particular are promising improved and faster deliveries, overnight and express deliveries as well as same-day deliveries. Intelligent solutions that are meant to facilitate quick, affordable and efficient deliveries to the customers have been researched into for some time now. Due to the strain on classical transportation routes, online retailers and logistics companies are now experimenting with the delivery by air with delivery drones. At present, the Deutsche Post lies ahead in comparison with Amazon and Google. In 2014, the DHL "Parcelcopter" started the first line operation with the first air transport for the carriage of emergency supplies with medications and urgent goods. The research project took off at the port in Norddeich and landed on the island of Juist on a special landing pad. An autopilot was developed for the smooth flight, which enables the automatic take-off and landing. The drone is said to be safe and robust in operation to cope with challenges such as wind and sea weather.

In contrast to the drone, the DHL "SmartTruck" has already been put into operation in Berlin. It is a delivery vehicle that is equipped with a new kind of tour-planning software and uses RFID technology. DHL gathers congestion alerts in cooperation with the Berlin taxi firms "if taxis are stuck in congestion anywhere in the German capital, the information detected by GPS automatically ends up at DHL. This is made possible by a system called 'Floating Car Data' (FCD), which was developed by the German Aerospace Centre".

At present, parcel deliveries without any driver whatsoever are being tested by robot suppliers. Some logistics companies, including the parcel service Hermes are testing robot delivery men for the suitability to deliver.

The company Starship Technologies has developed a driving robot delivery man. In cooperation with Hermes, the robot is meant to deliver parcels at the time chosen by the recipient. The electrically driven delivery man of 50 centimetres in height drives at walking pace on the pavement from the Hermes parcel shop to the customer. The recipient receives a code via a link with which they can track the parcel. They are informed about the arrival of the parcel via a text message sent to the mobile phone number given by them. The robot moves completely autonomously by capturing his surroundings and recognising hurdles such as traffic lights and zebra crossings. However, he is still monitored by an officer of head office who can intervene in the event of disruptions and can remote control the robot. Equipped with a GPS signal and an alarm, the parcel is said to be protected from thefts.

There is presently quite some research work going on on the basis of artificial intelligence in the area of outbound logistics. Until recently, it was difficult to apply intelligent robots in logistics as these processes comprise changeable and flexible activities. Innovative developments optimise logistic processes today, be it saving time during commissioning, reducing processing times or in supporting the employees in the core business. The error quota has decreased considerably, which leads to increasing effectiveness. Not only companies but also customers benefit from intelligent systems. This means the desired delivery time can be determined flexibly. Besides further factors such as terms for returns and delivery costs, fast and reliable deliveries lead to greater customer retention. For this reason especially, companies have to optimise their logistics processes and rely on intelligent systems. New developments appear to represent good alternatives, must, however, be well thought through. Currently, the new developments lack high safety standards. These challenges are to be mastered, and this is only a question of time. Companies should make use of the potential of artificial intelligence and robotics, in order to not miss the innovative transformation. In the future, it is to be expected that work in logistics will be given a fully new meaning.

3.6 Algorithmic Marketing

The times of not knowing which half of one's marketing budget works (Henry Ford) have for the most part become obsolete thanks to big data and AI. The following chapters will explain and illustrate this.

The automation of marketing processes has become common practice since about 2001 when collecting big data gained in importance. The data sets comprise, for example, customer databases or clickstream data which is a record of the customer's navigation between various websites. The amounts of data have, however, increased at a virtually explosive rate; this is how 90% of all data emerged in the twelve months prior to the beginning of 2016. As many companies do not know how they can use these data volumes with the former database systems and software solutions, the full potential of big data is not yet exploited by far. The traditional methods of automating marketing do not provide deep insights into the data either, do not foresee the effects of the measures and do not influence customers in real time.

If, however, algorithms are used for marketing, the data sets can be processed more efficiently. Algorithms can analyse and partition large data sets and recognise both patterns and trends. They can observe changes and recommendations for measures in real time, i.e. during the interaction with the customer. As well as that, thanks to the application of algorithms, marketers can dedicate themselves to more demanding tasks, which can result in a more efficient and more cost-effective marketing process. In the long run, due to the use of algorithms in marketing, companies can achieve a competitive advantage as well as a higher level of customer loyalty due to the greater customer proximity.

3.6.1 AI Marketing Matrix

Nowadays, there already is a multitude of potential applications for marketing based on artificial intelligence. These potentials can, in principle, be subdivided into the dimensions "automation" and "augment" as well as on the basis of the respectively associated business impact. In the case of the augment applications, it is especially a matter of intelligent support and enrichment of complex and creative marketing tasks that are currently still performed by human actors. Artificial intelligence can, for example, support the marketing team in media planning or in the generation of customer insights (see the practical example Sect. 5.8 "The Future of Media Planning—AI as a game changer"). First and foremost, the augment potential is already more strongly developed in those companies that reveal a high degree of maturity in the AI maturity model. Planning and decision-making processes are also supported or already performed here by artificial intelligence. With regard to the automation applications, it is hardly surprisingly noticeable that with them, both the degree of maturity and the distribution are significantly more developed in comparison. There are many automation applications, for example, that already have a high degree of maturity and use in practice today. This includes marketing automation or real-time bidding, for example (Fig. 3.11).

There are, however, applications that are used comparatively little in practice today despite their high degree of maturity and high business impact. One area of application this phenomenon applies to is the principle of lookalikes that can be used for lead prediction and audience profiling. In the B-to-C field, this can easily be put into practice with Facebook Audiences (https://www.facebook.com/business/a/custom-audiences).

This principle can also be easily applied in the B-to-B area (see practical example Sect. 5.1 "Sales and Marketing Reloaded—Deep Learning Facilitates New Ways of Winning Customers and Markets"). Behind this is the possibility of strategically identifying new potential customers on the basis of the best and most attractive key accounts of a company, who are similar to the key accounts in such a way that it can be presumed that they are likewise interested in the company's products.

The way it works is easy to understand: Customers—in the B2B area, these are companies—can be characterised on the basis of various aspects. Besides classical firmographics such as location, business sector and the company's turnover, these also include information about their development, digitality and their topical relevance. In times of big data, this enor-

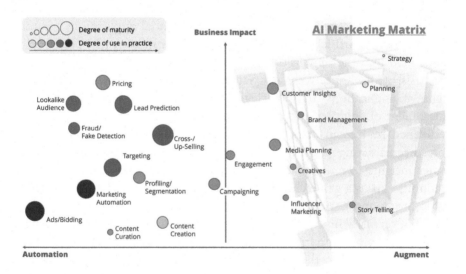

Fig. 3.11 AI marketing matrix (Gentsch)

mous amount of information can be mainly acquired from the companies' presences on the web, because every day, up-to-date posts about new products, changes within the company as well as on other subjects are published on the website and on social networks. On the basis of these aspects, all companies can be characterised comprehensively, on the basis of which a generic customer DNA is generated. In a subsequent step, further companies that have the same DNA—the so-called lookalikes—can be identified on the basis of this generated generic customer DNA. The result is a pool of potential new customers, the approaching of whom offers promising opportunities.

Thus, in the end, the conversion rate can be increased considerably in both marketing and in sales by using automated applications based on artificial intelligence. Practical examples reveal an increase in the conversion rate of up to 70 percentage. It is thus clearly becoming apparent that the principle of lead prediction and the identification of so-called lookalikes is an area of application with considerable potential and a great business impact for marketing and sales.

3.6.2 The Advantages of Algorithmic Marketing

- Efficient analysis of data sets
- Grouping of the data
- Recognition of patterns and trends
- Observation of changes in real time
- Reactions to changes in real time
- Efficient and cost-effective marketing process
- More time for creativity
- Long-term competitive advantage and a higher degree of customer loyalty
- Customer journey intelligence

On the basis of big data tracking, the "customer journey" can be systematically measured via different touchpoints such as search, social media and advertisements. On the basis of the data acquired in this way, media and marketing planning can be optimised with the help of so-called attribution modelling. From a multitude of data and points in time, the data mining model calculates the ideal channel mix by calculating the value proposition of each touchpoint in the overall channel concept. This way, which touchpoints have a direct conversion function and which have rather an assistance function can be accurately defined. Likewise, conclusions can be made about the temporal cause and effect chains.

It is interesting and important for companies to store customer data, in fact from the pre-acquisition phase to the conclusion of the customer relationship—in a manner of speaking the entire so-called customer journey. From the combination of this customer data with further factorisation information, with customer service aspects and other sales and marketing aspects, intelligent algorithms can make business decisions, derive recommendations for the businessman and conduct market research.

Even the customer journey to the purchase of a product provides strategically valuable information. This customer journey to making the decision to purchase is usually taken in several cycles, ideally in six steps: Identify need, research, receive offer, negotiate and purchase, after-sales and word-of-mouth communication. The touchpoints form the starting points where data such as tracking data or clickstreams is collected and analysed. This way, predictions can be made about future customer journey patterns. Networked points of contact can be prioritised in the scope of a digital strategy.

The advantage of this data- and analytics-driven approach is the empirical earthing. Data is neutral and objective and they make the same statement on Monday morning as on Friday just before going home. The digital "leaders" such as Apple, Google, Facebook and Amazon demonstrate how much company success is determined by data integrity, data quality and data diversity. The information is more topical, faster and more easily available than an annually recurring internal campaign "to better look after the CRM system again".

3.6.3 Data Protection and Data Integrity

As a matter of principle, when it comes to data protection, a differentiation must be made between personal data and data involving companies. As soon as inferences can be made to a specific individual and single data levels are being worked at, a moment has to be taken to consider: What is being processed? Is there already a business relationship? Which permissions or legal consent elements are at hand? Customer data may not be collected without permission and may also not be resold. Anybody who acts carelessly here can quickly render themselves liable to prosecution.

In principle, the following applies however: Almost anything is possible with the customer's consent. This is the reason why Facebook can act with the data to such an extent, because consent has been given, even if only few users have probably fully read and understood the Terms of Use. Likewise, a relatively far-reaching data processing in the scope of an ongoing customer relationship under the motto "for our own purposes" is possible and permitted. This could cover the likes of market research, acquisition activities and advertising.

In correlation with digitalisation, we frequently hear the keyword data integrity: It is in fact existential for businessmen as nobody can or wants to divulge more data on the Internet than absolutely necessary Data integrity means nothing other than knowing exactly what is happening to one's own data and to only share as much data as is actually necessary. This also includes critically reviewing the use of one's data and online services, portals and databases availed of—third-party providers, above all how they handle the entrusted company data. Data integrity thus means for businessmen who is allowed to find, use and disclose data and when and where.

The following chapters are initially dedicated to the use of algorithms in all four steps of the marketing process. Afterwards, practical examples as well as proposals for the right handling of algorithmic marketing will be given. The anticipated effects of algorithmic marketing on the economy as a whole will then be briefly presented.

3.6.4 Algorithms in the Marketing Process

Algorithms, e.g. in the shape of bots, can be applied in all four steps of the marketing process. In the situation analysis, in the marketing strategy, in the marketing mix decisions and in the implementation and control.

The situation analysis is meant to identify the customers' unfulfilled wishes. Bots can be applied in the internal situation analysis of identifying the key performance indicator that provides information about the company's strengths and weaknesses. In an external situation analysis, bots can search for certain keywords on the Internet to learn more about the customers and the competitors. Consumer behaviour can be observed and analysed with the help of bots. If companies use chatbots in customer service, bots can observe the courses of conversations and analyse them to obtain more information about the market and the customers. Bots can also hold interviews with certain customers or trend experts to conduct qualitative analyses. This can save both time and money as the interviews can take place at different places at the same time. Algorithms that can make predictions about factors and effects influencing the marketing activities (predictive modelling algorithms) can be used to research future demand.

In the second step of the marketing process, the creation of the marketing strategy, target groups can be identified with the help of bots that segment the amount of customers and analyse them according to various characteristics. The definition of the value proposition of the product, however, needs both creative and analytical skills, making this task less suitable for automation.

A widespread instrument for implementing strategic decisions is the marketing mix with the four Ps: Product, price, promotion and place. Algorithms can be applied in the following areas:

- **Product**: Chatbots can be applied in customer care, for example. Moreover, algorithms enable companies to develop new and innovative products and services that are tailor-made to the customer.
- **Price**: Product prices can be automatically changed with the help of algorithms, depending on the demand, availability and prices competitors have. Examples of companies that apply this dynamic pricing are airlines as well as Amazon and Uber.
- **Promotion**: Algorithms with AI can learn the customers' purchasing behaviour and needs and thus display individualised content and product recommendations to the customer. This is more efficient and cheaper than mass advertising for the company and can happen in real time. In addition, mature self-controlled recommendation systems can increase the opportunities of cross-selling, the offer and sales of additional prices.
- **Place**: Bots alleviate electronic commerce, also called e-commerce. If payment information and delivery address have been provided, the entire transaction can be performed by bots. On the basis of previous purchasing behaviour, a personal butler can also independently decide where a product will be bought. This can, however, also be problematic as this means the customer's purchasing behaviour can no longer be measured in the long term. The question is also posed as to how to proceed with regard to brand management in the future.

Many aspects in the last step of the marketing process, that of implementation and control, can be taken over by algorithms. Examples for the implementation of marketing strategies are, for example, the running of ads, the launching of a website or the sending of e-mails. As discussed previously, bots can display individualised Internet adverts. Bots can even take over the creation, personalisation and sending of marketing campaigns by e-mail. Even the creation of websites with the help of bots is possible, The Grid has been offering a private beta version for this since 2014 (Thomas 2016).

The control phase at the end of the marketing process can be performed in both a qualitative and quantitative way and is essential. Factors that should be controlled are, among others, the reach of the campaign, marketing budgets, customer satisfaction, market shares and sales. Algorithms can be helpful in this case to measure the various factors and to make statements about the efficiency of the campaign as well as to uncover potentials, such as increasing the customer lifetime value, of reducing customer acquisition costs. Apart from that, algorithms can improve the accuracy and efficiency of the control. The evaluation and presentation of the analyses data can be taken over by smart process automatisation software that is able to train itself or be trained. It can perform more complex and subjective tasks by recognising patterns. In addition, the data can be visually interpreted in the shape of dashboards.

3.6.5   Practical Examples

In some sectors, the use of algorithms is common practice such as in the production for controlling processes and in the financial sector for stock trading. In the recent past, it has also been shown that algorithmic marketing can increase a company's turnover.

3.6.5.1 Amazon

One example is Amazon that uses algorithms and that even grew in the recession. It is striking that the company has invested comparably high amounts in IT (5.3% of the sales revenue), whilst the competitors Target and Best Buy only spent 1.3% or 0.5% respectively. Amazon's dynamic pricing responds to competitor prices and current stocks. The investment in complex recommendation algorithms has automated 35% of the sales and 90% of customer support. This reduced the costs at Amazon by three to four percentage.

3.6.5.2 Otto Group

The Otto Group applies big data and AI for marketing and media controlling. On the basis of customer touchpoint tracking, a customer's activities can be systematically measured via various touchpoints such as search engines, social media and online advertising. With the help of the so-called attribution modelling, the Otto catalogue shop has optimised their media and marketing planning on the basis of the data acquired in this way. The model calculates the ideal mix of communication channels from a multitude of data and touchpoints by automatically identifying the value proposition—the attribution—of each touchpoint. This way, at which touchpoints the customer is directly animated to make a purchase can be accurately defined, i.e. which ones have a direct conversion function and which have rather an assistance function. The temporal cause and effect chains can be equally derived Otto systematically derives marketing measures and media budgets from this. The multitude of digital touchpoints and devices as well as their extremely variable use by the customer can no longer be optimised through experience and gut feeling. This empirical earthing ad objectification of marketing help to question the opinions and barriers that are frequently formed by the respective channel and contribute towards a significant increase in its effectiveness.

3.6.5.3 Bosch Siemens Haushaltsgeräte (B/S/H)

In order to obtain consumer reviews of products, classical market research avails of an extensive instrument. The significant disadvantage of this method is the effort associated with it. On the Internet, thousands of product reviews can be automatically analysed at any given time.

Seen systematically, this cannot be realised without big data. Ratings and reviews that are distributed across various Internet platforms need to be captured and integrated intelligently. In order to be able to quickly react to product reviews, this data also has to be captured fast, analysed and measures implemented. Companies can thus quickly respond to negative reviews. Positive reviews can be implemented in the marketing communication via websites, social presences or other product advertising means. BSH manages on the basis of a big data infrastructure as software as a service (SaaS) the entire process from the generation, capturing, analysis and use of the ratings and reviews. By way of these automatic rating and review analyses, customer reviews can be examined both in terms of quality and quantity and be used in a meaningful way for sustainable increases in turnover. BSH's internal analyses reveal, for example, that products with positive reviews achieve an increase in sales of up to 30%. These product rating and review analyses are thus becoming modern gold-diggers of the new Stiftung Warentest.

3.6.5.4 UPS

The logistics company UPS has also set themselves a target of saving up to US$ 400 million by using an algorithm that is meant to identify the most efficient transportation route. The taxi firm Uber uses an algorithm to bring together drivers and passengers. When a journey is requested, the algorithm offers the journey to a driver who is nearby. This equates to the so-called Supplier Pick Model, i.e. the provider selects. Similar to Amazon, the company uses a dynamic pricing system. If the demand for travel is high in a certain region, the price is increased by a certain factor that is known to the driver but not to the customer.

3.6.5.5 Netflix

Netflix, the online service for streaming films and TV series uses algorithmic marketing to personalise the content for the users and to recommend titles. A total of 800 developers work on the algorithms with the aim of keeping viewers. The social networks Facebook and Twitter as well as the online video channel YouTube use algorithms to select the posts that are displayed to the user. For Facebook, for example, the visibility of an (advertising) post is determined from various factors such as the popularity of the company's page, the success of past posts, the type of content (videos are preferred over photographs) and the time when the post was created.

3.6.5.6 Coca-Cola

There are, however, use cases of algorithms that demonstrate the dangers and limitations of algorithmic marketing. Coca-Cola, for example, has a Twitter account that converted negative tweets into cute ASCII images when they were marked with the hashtag #MakeItHappy. Subsequent to this, the US American magazine Gauker created a Twitter bot that published lines from Hitler's *Mein Kampf* and gave it the hashtag. Coca-Cola also converted these without further checking into images of dogs and palm trees.

3.6.5.7 Bank of America

The Bank of America operated a bot that was meant to help customers with complaints via Twitter. When an angry Occupy activist turned to the bank's

Twitter account, it sent the same prompt and standardised replies it sends for request for help from customers. The Bank of America ensured, however, that humans and not bots were behind the replies.

3.6.6 The Right Use of Algorithms in Marketing

As suggested by the afore-mentioned negative examples, certain risks are lurking in the background for companies that use algorithms in marketing. It is thus essential for companies to fully understand the algorithms applied and their limitations and for the algorithms to be used wisely. In addition, algorithms have to be supervised and controlled so that they are in harmony with the principles of the company and the image of the brand.

Another aspect is the ever-increasing concerns of customers regarding their privacy, which can arouse mistrust of the use of algorithms. If the customer sees too much personalised advertising, this can be perceived as creepy, especially if the advertising is based on very deep insights into private information. This is also called overkill targeting and can reduce the success of the marketing strategy, The creepiness that the customer can experience emerges from an imbalance in the distribution of the information. The company advertising knows more about the customer than the other way round.

Companies also need to be aware that by the collected and analysed data, they have an advantage over the customer and can thus manipulate and misguide their perception. If consumers are only shown pre-sorted information, they have no chance of obtaining an overall view. There is thus the risk that individuals exploit algorithmic marketing without heeding any ethical aspects. For the trust of the customer to be gained, the marketers must ensure that the algorithms adhere to the codex of digital ethics and privacy, and observe manipulation and selection of information as well as communication behaviour.

For a successful application of algorithms in marketing, it must also be considered that not all factors are analysed in context. The customer's mood, the weather or the presence of other people, for example, can influence the customer's purchasing behaviour. For this reason, an algorithm should contain as many variables as possible but also elements of surprise and chance, in order to not be too predictable. Another disadvantage of algorithms is that they are often restricted in their ability to analyse why a customer made a certain decision.

So that mistakes like those of the Bank of America are prevented, algorithms and bots should be applied with caution. Ideal is a combination of algorithms and real human interaction in customer contact. In this connection, two cases are differentiated: The touchpoint between customer and company is either by chance or the customer approaches the company with certain expectations. The first case refers to advertising campaigns or recommendations on websites where the customer can be positively surprised by the advert equating to their preferences. This can improve the value of the brand. The other way round, a customer who is not interested in the adverts ignores them without the value of the brand being damaged. If, however, the customer has certain expectations of the company such as direct means of contact regarding a complaint, the brand can be damaged if the expectations posed cannot be fulfilled by the company. On the contrary, the brand value can increase in the second case if a customer is satisfied. This does not necessarily mean that no algorithms can be applied in this case. It is, however, important that they rather act under human supervision and that humans can intervene in the process where necessary.

3.7   Algorithmic Market Research

3.7.1 Man Versus Machine

Artificial intelligence is also increasingly gaining ground in the field of market research. Some say it is "the death of traditional market research", other experts argue that it is "a chance to focus on what is essential and achieve real depth of research results". One thing is certain: If machines are to replace a human, if they are to be applied meaningfully in production, hospitals and households, they also have to learn and act through observation and experience.

In market research, computer-aided programs can analyse the entire data material faster and more thoroughly so that the human on the other side of the computer can concentrate on the important detailed questions—algorithms and AI thus entail a degree of market research liberalisation.

Programmatic market research allows for data-driven automated market research in the B-to-B sector. With this, companies can not only analyse their own data, but also market data, data of other companies, industry data and much more, and use the results. In practice, these are methods with which a computer makes decisions of which some input information is summarised to form an overall decision. Furthermore, AI systems are capable of learning and based on the results of previous decisions are able to adapt their decision logic. "Experience" is what you would call it in humans.

Nevertheless human intelligence is superior in certain areas, especially when the topic is not limited to a particular field, as is the case with a gaming computer, where programmed data is quasi only retrieved. Computers that can deal with the unforeseen that has not been programmed, for instance if the data collection method of a variable has changed and the system recognises this independently and looks for solutions, will come close to human intelligence. This kind of intelligence, however, is based on holistic knowledge about the world and will remain reserved for humans for some time to come.

It is the business of market research to capture and comprehend consumers' motivations. Ideally, the insights gained this way give marketing the opportunity to tailor services and products to customers' needs even better. The foundation of the whole trade is the idea of a subject acting autonomously and making decisions which can be justified and influenced. The more data is available for this purpose, the better. Meanwhile learning artificial systems are an indispensable aid to analyse huge data volumes and help with decisions.

3.7.2 Liberalisation of Market Research

Typically 80% of the time in market research is spent on time-consuming tasks such as sampling, data acquisition and analysis, leaving only 20% for decisive detailed questions. By means of innovative big data and AI processes, this process can be automated so that market researchers have more time for really value-adding activities such as the interpretation of the analysis results and to derive recommendations and actions. Tomorrow's market research will be oriented less towards samples and interviews but rather pursue a real-time census approach with automated analysis.

By its very nature, market research is an extremely data-driven industry. Market researchers have always collected, edited and analysed particular data and then dealt with the interpretation of this data. In today's fast-paced world, however, we are facing an enormous volume of data, we have already been juggling with zetta- or even yottabytes for quite some time. The global data volume is doubling every two years, resulting in a task man cannot cope with alone. Luckily, state-of-the-art technology not only provides memory space and the adequate computing power to be able to deal with the mass of data, but also diverse evaluation and analysis possibilities. The latest developments in the area of machine learning allow making smart data from big data and using data really economically.

Successful market research has to adapt accordingly and integrate these innovations in its work if it does not want to be left behind. For example, there is already software which automatically converts the answers of subjects from studies (CAWI, CATI and CAPI) into codes whilst not only considering the respective main statements but also extracting and semantically linking all the other information. The significance is increased by a multiple thereof. Far-reaching interpretation then follows hereby the code plans reach a new level of detail difficult to achieve with manual processes.

But actually it is not about choosing either man or machine. AI systems are an intelligence amplifier. Poorly drawn up, poorly maintained and poorly interpreted, they only produce costs, trouble and nonsense. Well-programmed, capable of learning and used intelligently, artificial intelligence can save a lot of work and create time for depth of detail. When it comes to decision logic, for example, artificial systems are always more complex and by far more precise. And that is exactly why predictive analytics—i.e. the prediction of customer losses, of sales figures or price acceptance—is so useful. Also concerning the question "What causes the customer behaviour", i.e. a causal analysis, AI systems are considerably better. Because humans can actually only think in correlations and thus fall into the trap of spurious correlations on a regular basis, human decision-makers also have to learn something new.

In a first step, market research with artificial intelligence can complement the classic path, in a second step, however, in part even replace it. The digital index of the state government of Rhineland Palatinate compiled for the first time in 2015 is one example. ZIRP—Zukunftsinitiative Rheinland-Pfalz (initiative for the future of Rhineland Palatinate)—had polled 260 of 170,000 companies in the state beforehand, which was not only cumbersome, but also time-consuming and costly. In contrast, software based on artificial intelligence can provide information on 110,000 companies in an instant.

3.7.3 New Challenges for Market Researchers

Some market researchers tend to view the automation trend very critically. They are rightly proud of the traditional methods which have been undergoing improvements for decades and are based on the wealth of experience from the entire sector. The concern that automation means compromising on quality is not unreasonable. If artificial intelligence cuts out humans from market research, will not a lot fall by the wayside? Not necessarily, because there is little probability that AI will carry out market research without man in the future. Only us humans are able to really consider all contexts such as emotions, cultural influences, the small, yet significant differences. Here, artificial intelligence clearly reaches its limits. When selecting the data collected and interpreting it in a target-oriented way, man will continue to play a decisive role.

Whilst the automatic analysis of data recognises the behavioural patterns and characteristics due to the wealth of information, it is the task of modern market research to reason these behaviours from customers' attitudes and opinions. This certainly confronts market researchers with new challenges, automated processes cannot be applied without planning and testing. User-friendly applications that facilitate use are on the rise. With them, valuable time and money can be saved to focus on the essentials: Asking the right questions that lead to understanding the customer better thus enabling even better strategic decisions. Therefore, this is a complex field which is by no means satisfied by using a machine instead of a man.

Machine learning can, however, support good and creative research design. In traditional market research, questions have always been tackled with a specific hypothesis. These initial ideas are also considered in the evaluation and interpretation of data. Thereby, sight can be lost of the useful results one would not have expected in the beginning. Machines on the contrary have no prejudices and draw conclusions without bias. They are able to accurately evaluate a wide variety of information and to recognise unexpected events. This is where market researchers come in and can creatively continue working with the additional findings, plan new strategies and refine the design of their studies. Used correctly, machine learning eases the workload of market researchers so they stay focused on the broader picture.

Programming itself has to be "intelligent" too: Causal analysis, for example, cannot be standardised fully. It has to be drawn up, interpreted and maintained individually for each problem. Just like other AI systems—only to a higher degree—"truly" intelligent humans are required, somebody who knows what the causal model represents in terms of content, what the data means and how it was measured. If data is simply analysed without knowing what it means, how it was collected and how the data is analysed, this could result in wrong interpretations. In summary, it can be said that true intelligence develops through the simultaneous and holistic knowledge of facts and analysis method.

3.8   New Business Models Through Algorithmics and AI

Besides designing and optimising corporate functions and processes, algorithmics and AI also have the potential to challenge and reinvent business models.

Netflix, for example, owes its current success to the fundamental disruption of a business model from video-on-demand to streaming media. This way the company made it from the lower end of the market to the global market leader in no time at all, even ahead of the streaming portal of the giant Amazon. In addition to the production of their own series, Netflix won recognition in particular through the AI they developed, guaranteeing maximum, dynamically adapted streaming quality, even if the Internet bandwidth is very low. As a result, the company was able to even prevail on markets with a rather underdeveloped infrastructure and establish itself at the top.

Likewise, the agile start-up Airbnb is already threatening traditional industry leaders such as Marriott. Having started out as an idea of an inexpensive solution for budget travellers with air mattresses in living rooms of strangers, luxurious apartments can now be booked via the portal. The innovative price formation algorithm is setting new standards.

With the aid of a trained deep learning network, factors such as location, furnishings, demand, but also presentation are weighted differently in real time, and the system calculates a price tip for the host. In this way, the provider manages to serve an entire sector and offer all users the best price. The popularity of this business model speaks for itself!

The financial service provider sector is also positioning itself anew. Recently, the expanding and critically discussed start-up Kredittech has been stirring up the market—on the basis of big data, it calculates a consumer's creditworthiness score with a precision and term that would not be imaginable with conventional methods. This way creditors can minimise their risk of payment defaults and the credit applications of customers are accepted or rejected much faster (Fig. 3.12).

The B2B sector is also reacting with appropriate AI-as-a-service, allowing a digital, synthetic credit score to be calculated automatically on the basis of big data and AI.

Offers of Robo Advisor such as scalable capital in investment consulting or the clark.de app in insurance consulting and administration are also booming. Customers are informed about most recent market developments

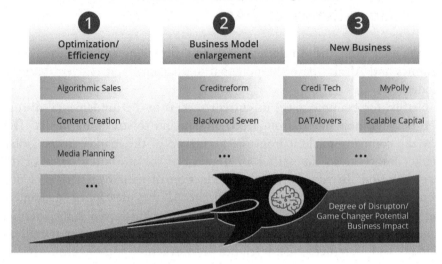

Fig. 3.12 AI enabled businesses: Different levels of impact (Gentsch)

in real time and are able to react. The offer can be adapted to exactly meet the customer's needs, and accessibility of offerings via mobile smartphone apps or Internet portals is not comparable with a local adviser.

In data economics, data also plays a central role as a source of expanded or new business models. Figure 3.13 provides a list of questions to determine the potential of data for expanded and new business models.

Considerations are to be made as to whether available data can be used to expand the business model or can be monetised through the sale to other companies. On the other hand, in line with the assessment of potential threats, it is necessary to examine whether competitors might possess data that pose a threat to one's own business model.

3.9   Who's in Charge

Do companies need a Chief Artificial Intelligence Officer (CAIO)?

In January 2018, Facebook created the new title "Vice President of Artificial Intelligence". With this, the largest social network enhanced the research and application of AI in January 2018. Facebook employed Jérôme Pesenti, who will be heading this area in the future. For years, Pesenti has been an established key figure in the industry. He created the Watson supercomputer offers for IBM and changed to the British artificial intelligence company Benevolent AI in 2016.

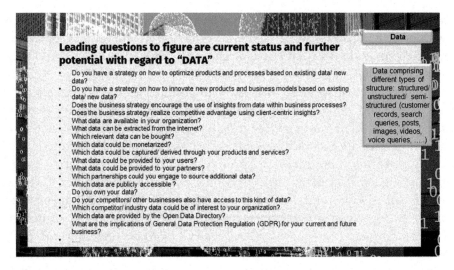

Fig. 3.13 List of questions to determine the potential of data for expanded and new business models (Gentsch)

Since 2018, he has been the head of the research department, which also deals with fundamentals, and additionally the group that attends to applications for machine learning. The new team composition signalises that Facebook is placing even stronger emphasis on the advances in artificial intelligence, which is used in more of the company's products, not only in personal timelines.

Whether "Vice President of Artificial Intelligence" or "Chief Artificial Intelligence Officer (CAIO)"—what is the motivation and rationale behind such a position?

3.9.1 Motivation and Rationale

The relevance of digital transformation for companies is uncontested across all industries and company sizes. The implications differ in urgency—length of the fuse cord—and in the degree of disruption—force of the blast. The central drivers and enablers in this respect are often data, algorithms and artificial intelligence. Frequently calls grow loud for a data scientist as a solution. This position and these skills are important, of course, but there is a lack of consolidation at managerial level. When a Chief Digital Officer (CDO) is recruited, they are often required to assume co-responsibility for this part. The question arises as to whether a new executive position has to be created based on the business relevance and complexity of the data and analytics issues in order to provide a technically specialised contact point within the company's own structure for certain tasks and business processes. Do companies need a CAIO in addition to a CDO, or does the CAIO even replace the CDO?

Traditional management and marketing strategies are too sluggish to act agilely and time-effectively. Decision cycles take too long because the structures are too rigid to use the insights gained according to the new paradigm of data-driven real-time business. An organisational structure is necessary that puts companies in a position to quickly and efficiently react to the requirements of digital transition. The area of marketing, for example, which has always controlled customer communication and implemented the sales objectives, is predestined to assume the leadership role in designing the transformation process. Usually the Chief Marketing Officer (CMO) is entrusted with this task, but it often proves to be too complex because horizontal and cross-departmental actions are required. Companies with a higher level of maturity therefore assign the CDO described above to management who is responsible for the digital transformation for the entire company and coordinates the interface to marketing.

Interlinking and optimising existing operational processes with new digital elements such as machine learning, algorithms and artificial intelligence represent a smart possibility to exploit the company's own potential. At the same time they confront tried and tested business models with new challenges of utilising the own data-driven potential in a strategically optimal way. Most companies still avoid internal recruitment concerned exclusively with the technical implementation of digital transition. For the most part, positions such as the CEO or Chief Information Officer (CIO) take responsibility for digital transition in general. More rarely the task is shifted to IT or marketing; there have been no clear assignments for the rest. Those who have already concerned themselves with this topic in more detail might consider the much debated position of the CDO that is becoming increasingly indispensable as a comprehensive contact point for the structured digitalisation of companies. This person purposefully leads the company into the required direction of digital transformation, promotes changes in the dialogue as connecting link between the levels relevant for the decision, and guides them through a technically trained organisation, as well as an assessment and exploitation of potentials.

Therefore, there is truly a need for a CAIO for companies who want to become and stay capable of acting digitally in the future. Besides this interdisciplinary position, the increased use of artificial intelligence and machine learning gives rise to the question of whether recruiting an in-house

CAIO—Chief Artificial Intelligence Officer—is also necessary for this specific technological area in order to further extend one's own competitive advantage.

3.9.2 Fields of Activity and Qualifications of a CAIO

To develop a clear answer it should first be defined which scope of duties the position of a CAIO is in charge of and what professional requirements the potential candidates have to fulfil. Finding themselves confronted with the need to switch to electricity in former times as a transformation process to stay competitive, the digital turnaround nowadays is subjecting the standards for recruiting to its own rules. Apart from a minimum of ten years of industry-specific professional experience, where he was able to develop within his own success and also failure story and was able to learn to coordinate himself as a team player, the post of a CAIO especially demands direct experience in the technical areas of data analysis, cloud computing and machine learning.

In close coordination with the Chief Technology Officer (CTO) and CIO, it will be his specific task to develop innovative digital approaches to a solution by means of the existing range of products whilst at the same time promoting the use of machine learning across the company. For this, it is necessary to figure out the internal strengths and weaknesses in order to subsequently work out solutions for a specific company-related AI strategy. In addition, it is his task to set up new partnerships focused on artificial intelligence and to find relevant platforms in order to finally optimise customer satisfaction and product choice by using the developed AI improvements throughout the company.

By means of targeted strategies, a CAIO is able to expand unutilised data silos to a sustainable competitive advantage with the aid of machine learning and to reduce financial expenditure in close cooperation with the cost centres. Personal abilities such as natural leadership skills, consolidated through corresponding managerial positions in large-scale enterprises, but also experience in working with start-ups are just as essential as technical know-how, evidenced by a focus on an AI-oriented work routine in the field of machine learning, data analysis and assessment. The profile is rounded off by several years of experience in creating or advancing and implementing data-driven solutions for products and platforms accomplished in the area of focus through machine learning or cloud computing. An educational background in the IT sector is also of great advantage. In summary, the ideal candidate has experience in both "tangible" and digital competition and is character-ised by his ability to work in a team and his personal initiative as well as by handling data-driven applications in a solution-oriented and innovative way in the field where he possesses longstanding profound experience and excel-lent expertise.

3.9.3 Role in the Scope of Digital Transformation

In the scope of the digital transformation, this highly qualified CAIO is able to further promote the digital transition in close cooperation with the rest of the AI team and to identify strategic competitive advantages, which reflect in the company as sustainable growth in value. In the long run the suc-cess of his role will develop in parallel to the technical advances of artificial intelligence and grow together with them. Due to the fact that data-driven processes will increasingly play a major role in competition, the call for cor-responding qualified staff is steadily getting louder. And just as the area of responsibility of AI within the company is noticeably expanding, the range of tasks of the CAIO will be subject to continuous extension in the course of the digital revolution as was once the case with the transition to electric-ity. In the process, the CAIO will specifically adapt and further expand the already existing knowledge about AI applications and therefore be able to get the maximum benefit in a company-related context. In the course of the coming years, these developments—driven by digital transformation itself—will both increase in speed and innovation and thereby at the same time gain more and more central importance within the company.

3.9.4 Pros and Cons

But this outlook on the requirements and development potential of a CAIO is followed by the question whether the investment of capital and personnel in the creation of such a position proves to be just as indispensa-ble as that in favour of a CDO. The integration of intelligent data systems not only offers future-oriented companies advantages, but also confronts them with internal challenges of developing and implementing the right AI strategies. Decision-makers face the question: What are the minimum requirements for an AI team at this stage of the digital transformation and what positions are indispensable or superfluous? A CAIO as a catalyst for the enormous amount of data within the value chain can be absolutely profitable. And if occupied carefully, such a position brings a huge benefit so that such an investment does seem valuable. But the digital revolution is only just getting warmed up.

Both the corporate landscape and consumers first have to fully embrace notion of a digitalisation of the market to be able to benefit from it in an ideal way. Before all too detailed AI strategies can be developed, it is first necessary to have a solid basic understanding of the market needs in its present state as well as the underlying data-driven potential for change. The driver of a transformation is thus not the digitisation per se, but rather the company objectives themselves. Only with an individual and result-oriented approach can the maximum potential from the data pool be utilised. This requires interface positions such as the CDO and the Chief Data Scientist, who fundamentally deal with the analysis and integration of digitalisation strategies. Only when companies have undergone this initial and fundamental change can positions like that of the CAIO be considered. In an already digitalised company, he will disclose the possibilities for AI applications as well as their maximum benefits. But a technically highly qualified position will not be able to solve the initial problems which such a digital disruption involves.

A better strategy is to first identify the problems involved in digital change, and develop digital strategies in a solution-oriented way, before taking the next step towards innovative optimisation. To do so, the AI team has to be established and integrated in-house and the tasks that can be improved or assumed by AI have to be identified in a first step in a constant dialogue at different levels and in a last step, approached from a technical point of view. Only when the processes operate well together and AI generates a benefit as an integral part of the company, should fine-tuning be considered. The CAIO can further optimise this development process, but he needs an experienced team and contact points who provide him with the cores of the problems for which he is supposed to develop technical AI strategies.

3.10 Conclusion

The skills and tasks of a potential CAIO are indeed important for a successful digital transformation and for optimising existing business models. Ideally, however, his tasks should be covered by the role of the CDO in conjunction and connection to the Chief Data Scientist. Companies run the risk of handling the executive label inflationary and establishing non-synchronised shadow organisations. Conclusion: executive relevance of algorithmics and AI: Yes—own executive position: No.

References

Mckinsey. (2017). http://www.mckinsey.com/business-functions/digital-mckinsey/our-insights/intelligent-process-automation-the-engine-at-the-core-of-the-next-generation-operating-model. Accessed Mar 2017.

Mitchell, T. M. (1997, March 1). *Machine Learning* (1st ed.). Blacklick, OH: McGraw-Hill Education.

Russell, S. J., & Norvig, P. (2012/2016). *Artificial Intelligence—A Modern Approach.* Upper Saddle River, NJ: Pearson Education.

Thomas, T. (2016). *Artificial Intelligence in Digital Marketing: How Can It Make Your Life Easier?* http://boomtrain.com/artifcial-intelligence-in-digital-marketing/. Accessed 4 Jan 2017.

Turing, A. (1948). *Intelligent Machinery* (p. 1982). Berlin: Springer.

Part III

Conversational AI: How (Chat)Bots Will Reshape the Digital Experience

4

Conversational AI: How (Chat)Bots Will Reshape the Digital Experience

4.1 Bots as a New Customer Interface and Operating System

4.1.1 (Chat)Bots: Not a New Subject—What Is New?

Bot, find me the best price on that CD, get flowers for my mom, keep me posted on the latest developments in Mozambique.

—Andrew Leonard (1996)

The topic of bots is new. Back in 1966, Joseph Weizenbaum developed with ELIZA a computer program that demonstrated the possibilities of communication between a human and a computer via natural language. When replying, the machine took on the role of a psychotherapist, worked on the basis of a structured dictionary and looked for keywords in the entered text. Even if this bot model as a psychotherapist only celebrated questionable success, such bots of the first generation with a firmly predefined direction of dialogue and keyword controlled are still used in many places.

Especially in the past two years, bots have been experiencing a new quality and significance due to the fast developments of artificial intelligence, platforms, communication devices and speech recognition so that the unfulfilled wish of Andrew Leonard in 1966 can finally become reality.

Communication and interaction are increasingly controlled and determined via algorithms. Bots and messaging systems are being hotly debated and frequently have to serve as the mega trends of the years to come. The

© The Author(s) 2019
P. Gentsch, *AI in Marketing, Sales and Service*,
https://doi.org/10.1007/978-3-319-89957-2_4

focus is primarily on communication interfaces that bring along efficiency and convenience advantages as the next logical level of evolution. But it is about way more than "Alex, order me a pizza please" or "Dear service bot, how can I change my flight?"

The popularity of messaging and bot systems is increasing constantly. Since 2015, more people have been using applications (apps) for communications than social networks. That is almost three billion people worldwide every day. In Europa and in the USA, the platforms WhatsApp (approx. one billion people) and Facebook Messenger (900 million) are mainly used, whereby in Asia, WeChat (700 million) and Line (215 million) dominate.

Two of the most significant companies of today, Microsoft and Facebook, announced in the spring of 2016 that will be focusing on bots in the future. Microsoft, whose CEO Satya Nadella describes bots as "the next big thing", is said to be fully concentrated on the company-own personal assistant Cortana in 2020 according to an analysis by the IT research institute Gartner. Instead of the current heavyweight Windows, robots and chat platforms are to move into the focus of Microsoft's strategy. All in all, the Gartner Institute expects that in 2020, 40 percentage of all mobile interactions will be controlled by bots (Gartner 2015).

4.1.2 Imitation of Human Conversation

At the beginning, bots were able to answer simple, repetitive questions that follow simple rules such as "What is the weather like today?" With the advances in artificial intelligence and machine learning, bots can now take over more demanding tasks. The idea of the bot goes back to the 1950s when Alan Turing, a former researcher in computer intelligence, presented a test to test the intelligence of machines. This is known to this day as the Turing test and works as follows: If more than 30% of an experimental group are convinced that they are having a conversation with a human and not with a computer, an intellectual power on a par with that of humans is assumed of the machine.

In 2014, there was a small breakthrough in this respect when a third of the participants were convinced that they had been having a conversation with a human, although a bot had been used. It is not always easy these days to see the difference between a human and a machine in a conversation. Comparably little artificial intelligence can suffice to imitate the illusion of a natural human interaction. The developers of bots, however, still face many challenges in this respect. Their aim is to develop a common language between machine and man to alleviate communication.

4.1.3 Interfaces for Companies

So that companies can offer their services on messaging platforms, there have to be application programming interfaces (API). The APIs allow the integration of an external programming code, like a bot, in existing software, for example a messaging platform.

Not all companies have the expertise for building their own bots and integrating it into a messaging platform. It is thus probable that, in the future, there will be increased numbers of bot-as-a-service concepts, simplifying the development and integration of bots. Sara Downey (2016), director of a start-up investor, thinks that the developed bots should be both universal and simple to build. Universal means that the bots are to be easily maintainable on all different kinds of platforms. And if it I simple to build a bot on top of that, not only could the tech experts of the company could be assigned the task but also employees with a talent for language and communication. Two such bot builders are already available via Facebook and Microsoft and will be presented in the following paragraphs.

In April 2016, at the annual Facebook developer conference F8, the company reported that it had created new interfaces for Messenger for external developers. Wit.ai, software that helps to develop an API for speech-activated user interfaces, was connected to Facebook beforehand to alleviate developers in the integration of their services. In the first two and a half months after Messenger was launched, more than 23,000 developers have registered on wit.ai and more than 11,000 have emerged. In the meantime, Messenger also offers a visual user interface to improve the user experience and contains plug-ins that can integrate the bot in offers by third-party providers. Since autumn 2016, it has also been possible to effect payments directly via Messenger. If the credit card information is stored in Facebook or Messenger, the transaction can be concluded without further entries. Many companies have already connected with Facebook Messenger. One example from Germany is bild.de, which operates a live ticker via Messenger.

The Microsoft bot framework creates the conditions for developing bots for various platforms or one's own website. The bot builder software development AIt (SDK) enables the bots to be implemented. The Language Understand Intelligence Service (LUIS) assists the bot with deep learning and linguistic analytics. The bots can be integrated in various messaging platforms with the bot connector The bot directory facilitates the distribution and discovery of other bots in the platform.

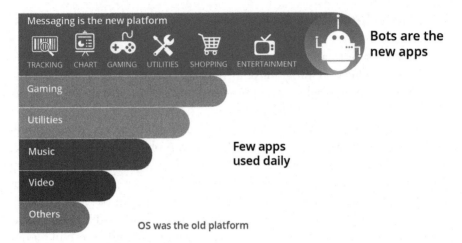

Fig. 4.1 Bots are the next apps (Gentsch)

In January 2016, WhatsApp, which also belongs to Facebook, also announced that they want to test tools that realise communication with companies. Further examples of platforms that allow the building and integration of bots are Slack, Telegram and AIk.

The development of bots will lead to fundamentally different principles in communication and in the corresponding interfaces. Bots will replace the majority of websites and apps. The separation of application-related functions. A transaction can, for example, contain the evaluation of a product, the selection as well as the purchase and service. Typically, a consumer would have to use different apps and/or websites for this. The bot as a kind of operating system combines various forms of information and interaction to become a universal transaction (Fig. 4.1).

The bot has made a selection as per the references learned, triggered the order and completed the transaction using the bank and address details known to him. Of course, appropriate permission states are integrated and which are controlled by the respective consumer.

4.1.4 Bots Meet AI—How Intelligent Are Bots Really?

Chatbots are currently being boosted with the performance attribute AI. However, most bots at present are being implemented in a relatively trivial way. As a rule, certain keywords are scanned for on Twitter and Facebook, on the basis of which predefined texts or text modules are then automatically played out. Somewhat more intelligent are systems that automatically detect relevant text findings on the Internet and then put them together accordingly to form a post.

This automatic form of content curation is also discussed under the term robot journalism. For the chatbots to be able to capture the posts accordingly, the in the meantime significantly advanced processes of natural language processing (NLP) which transform the running text into corresponding semantics and signal words, are used.

Another approach is to connect the chatbots to knowledge databases. To the user, chatbots seem to be "intelligent" due to their informative skills. However, chatbots are only as intelligent as the underlying database.

Due to the advances in AI, chatbots can be by all means made more intelligent in the future. AI-based chatbots learn largely independently from the huge amounts of data available online and recognise question-and-answer patterns that they use automatically in customer communication. The example of Microsoft Tay mentioned shows, however, that the uncontrolled training of the bots by the community can lead to fatal consequences. The next generation of AI-based bots must control and create the possible room for communication.

With that, the degree of information supply is directly associated with the degree of intelligence and automation of the bot. The present-day (usually unintelligent) chatbots are fed the keywords, knowledge modules, texts and rules of their developers/programmers. The more intelligent form of bots obtains this information themselves from online sources and combines it to form new content. The AI-based bots are also fed by the answers and reactions of the users. The possibility of controlling thus also sinks for the information used for learning.

Important food besides contents is also social signals such as likes and followers. These enhance or reduce the impact of chatbots. This feedback information can also come from other bots. So-called bot armies can make contents and opinions go viral within a short time and thus automatically set topic and agenda trends.

At the beginning, bots were able to answer simple, repetitive questions that follow simple rules such as "What is the weather like today?" With the advances in artificial intelligence and machine learning, bots can now take over more demanding tasks. The idea of the bot goes back to the 1950s when Alan Turing, a former researcher in computer intelligence, presented a test to test the intelligence of machines. This is known to this day as the Turing test and works as follows: If more than 30% of an experimental group are convinced that they are having a conversation with a human and not with a computer, an intellectual power on a par with that of humans is assumed of the machine.

Eugene Goostman, a chatbot that has been developed since 2001, is said to have succeeded in this. The bot imitated the personality of a 13-year old Ukrainian boy. At a competition in 2014 that was organised for the 60th anniversary of the death of Alan Turing, Eugene Goostman succeeded in convincing 33 percentage of his human chat partners that he was human and not an AI system. Hereupon it was declared that the bot had passed the Turing test. This conclusion was, however, discussed controversially as it was seen as a trick to choose the character of a 13-year old Ukrainian boy, who could easily be misleading over gaps in knowledge and structural shortcomings.

To date, bots have been programmed quite trivially in the main, one could also say "dumb". In the times of artificial intelligence, this will change sustainably. Past implementations draw on internal databases, text modules tagged by keywords and rules of the developer. The bot scans the customer input, for example, for keywords, then compiles the knowledge and text modules according to firmly implemented rules and gives the output generated in this way back to the customer. Expansions of the system in the shape of new knowledge, rule combinations, keyword tagging and text modules have to be programmed. Questions the systems do not understand or to which they have no answer in their base are replied to with counter-questions and evasive manoeuvres.

Present-day bots use in addition the largest available dynamic database of the world. The semantic web, i.e. the collective effort of content uploaders to tag in the hypertext the information semantically and standardise and to thus make it machine-readable alleviates the automatic access to knowledge.

Via the customer's interaction patterns, the bot can find customer-specific keywords with the deep learning algorithms of machine learning and maintain its own database customer-specifically and automatically. The intervention of the developer is only necessary for maintenance purposes.

Current breakthroughs in NLP, the sub-area of AI that occupies itself with man-machine communication, increase the dynamism of the bot development even further. Back in 2014, chatbots that faked being a human vis-à-vis a third of the human users were successfully developed. In the meantime, it is possible in verbal communications to pull together 90% of the spoken word into context. Written communication in this field is, however, much more developed and thus more widespread.

Bob and Alice, two AI-based chatbots in Facebook's research laboratory for artificial intelligence, FAIR, invent a language that their human inventors do not understand. The original idea was to teach the chatbots how to negotiate. With that, the systems developed their own language between each other that not even the creators were able to understand.

This sounds sort of like this:

Bob: I can can I I everything else
Alice: Balls have 0 to me to me to me to me to me to me to me to me to

This independence and apparent loss of control was discussed in the press almost with panic down to apocalyptic end-of-the-world scenarios. Some saw the development to become Skynet; others the end of our civilisation in the spirit of super intelligence or singularity.

Yet, it is not quite as dramatic by far. Bob and Alice were to negotiate various items whereby certain items were more important to each bot than others were. The AI was meant to find out in dialogue where the other bot's preferences were. What did in principle work well, if the developers had not forgotten to reward the bots for following the modalities and rules of the English language. So Alice and Bob began to use a kind of computerised stenography. Facebook subsequently reset the bots to adapt the rewards system accordingly.

Much more interesting but not reported in the public domain is the fact that the systems that quasi additionally learned how to behave tactically and also to lie where necessary to get what they wanted to have. Alice and Bob behaved as if they were apparently interested in certain things to then leave them to the other bot. although the bots were not taught to do this, Alice ad bob were able to haggle over the item they actually wanted.

The chief AI developer of Facebook, Dhruv Batra, puts the scaremongering and the alleged loss of control into perspective: "Changing the parameters of an experiment is not the same as pulling the plug of an AI system. If that were the case, every researcher would be doing it constantly when a machine is meant to perform a different task".

In order to be able to apply such bot systems more purposefully in the future the FAIR researchers reinstalled the system after that with the objective of Alice and Bob being able to successfully negotiate with humans in the future.

4.1.5 Mitsuku as Best Practice AI-Based Bot

The bot Mitsuku, which runs on Pandorabots, one of the most powerful conversational artificial intelligence chatbot platforms, won the Loeber Prize in 2013, 2016 and 2017 for the world's most humanlike chatbot. Mitsuku answers extremely quickly and quick-wittedly so that one is under the impression for quite a whilst that one is speaking to a real person.

Figure 4.4 makes it clear that the quality of AI-based bots strongly correlate with bi data. As people from all over the world speak to Mitsuku, an abundance of global training is developed that enables the AI to learn and become permanently better. Even if Mitsuku was not developed for any specific company purpose, it shows well the quality that future bots will achieve on the basis of big data and AI.

4.1.6 Possible Limitations of AI-Based Bots

The examples above already show the present-day potential of AI-based bots. At present, these systems are still in an early stage and still have certain limitations and potentials for optimisation.

4.1.7 Twitter Bot Tay by Microsoft

Most bots at present are reactive service bots. Engagement bots that actively interact with the users as market and brand ambassadors go one step further. The most famous example here is the chatbot Tay by Microsoft.

Microsoft removed Tay from the web apologetically within one day. The example shows that the uncontrolled training of bots by the community can lead to fatal consequences. AI systems still have to learn ethical standards. It thus becomes apparent that even bots require a kind of guideline. Like a journalist has to observe editorial guidelines, bots have to observe certain standards. The next generation of AI-based bots must control and create the possible room for communication.

IBM Watson has been able to celebrate quite a few respectable results in the field of AI, such as winning the much-quoted Jeopardy game Champs Of The Champions (all winners of Jeopardy competed against each other). To make the system seem more human, the IBM researchers tried to add the Urban Directory as training database. The Urban Directory contains colloquial language and slang.

The limitations of present-day AI are evident in the fact that the system cannot really differentiate between obscenity and courtesy. Watson, for example, replied to a serious question a scientist asked with the word "bullshit" that was certainly not adequate in this context. Humans are able to intuitively conduct this interpretation and reasoning—present-day AI systems cannot.

(Chat)Bots as enablers of Conversational Commerce.

4.2   Conversational Commerce

In this chapter, the benefit and application scenarios of so-called Conversational Commerce are illustrated. It will be possible to see how Conversational Commerce through intelligent automation enables the optimisation of customer interaction. In addition, with the DM3 model, a systematic process model is presented with which the complex task of Conversational Commerce, which contains strategic, organisational and technological tasks, can be successfully implemented. Additionally, new trends and the consequences of these developments for companies will be described. The advantages and disadvantages that could result for consumers will equally be illustrated—keyword "personal butler" ("digital butler").

On the one hand, the digital transformation is driven by the technological developments and innovations, on the other hand, the increasingly smart and empowered consumer is becoming the driver increasingly more. In relation to e-commerce, it is technologies such as messaging systems, marketing automation, AI, big data and bots that are facilitating a transformation of existing e-commerce systems towards a higher degree of maturity in the spirit of the algorithmic business. On the other hand, the networked and informed consumer forces a real-time company to (re-)act fast and competently. E-commerce thus not only faces the question as to whether it has to change but rather how it must change. These two lies of development are currently being discussed under the term "Conversational Commerce".

Yet this does not mean automation and real-time messaging at any cost; in fact, how and when which touchpoints of the customer journey that should be automated and supported under cost-benefit aspects must be systematically reviewed, the benefit and application scenarios of Conversational Commerce will be illustrated in the following sections. In addition, with the DM3 model, a systematic process model is presented with which the complex task of Conversational Commerce, which contains strategic, organisational and technological tasks, can be successfully implemented.

4.2.1   Motivation and Development

To date, customers who wanted to contact a company had to either fill in forms or call hotlines, often with long waiting times on hold. This kind of communication can, however, often be one-sided, annoying and slow for the customer. On the other hand, communication with friends, acquaintances and colleagues is increasingly taking place via messaging platforms such as

WhatsApp or Facebook Messenger. We can now observe a departure into a new communication paradigm where companies are using messaging platforms, chatbots and algorithms both for the interaction with customers and for internal communication. This is mainly promoted by the advances in artificial intelligence that facilitate the creation of adaptive algorithms and chatbots that can automate communication whilst they still feel like humans.

The new communication paradigm brings along many trends such as Conversational Commerce (customer advice and purchase via conversation), personal butlers (digital personal assistants that take over the purchases, bookings and planning for the user), algorithmic marketing (integration of algorithms ad bots in all steps of the marketing process) and conversational office (integration of messaging platforms combined with bots in internal company processes). In this chapter, these new trends will be described and the consequences of these developments for companies will be illustrated. Quickly catching up with the new communication paradigms can, on the one hand, result in more efficient work processes, greater customer retention, an increase in sales and in a competitive advantage for companies. For customers, the increase in comfort is particularly crucial as annoying tasks can be seen to within minutes. If companies sleep through the trend, it may be the case that they are not taken into consideration in the selection of services or products in the future. On the other hand, many risks are lurking for companies as customers are much more easily disappointed and brands can be damaged more easily. Apart from that, the increased use of bots and algorithms can lead to job losses. For companies it is thus essential to understand the new trends and to know their risks.

4.2.2 Messaging-Based Communication Is Exploding

4.2.2.1 But Why Is Messaging Booming in Comparison with Other Apps?

Van Doorn and Duivestein (2016) from the SogetiLabs, a research network for technology diagnosed an app fatigue among users. As a matter of fact, only a very limited number of apps are used by each user every day. This could be due to the app jungle consumers are being confronted with. Rh frequently heard sentence "there's an app for that" not only appears to be true but also an understatement. Consumers are confronted with at least a dozen apps for every conceivable area of application. This makes it difficult

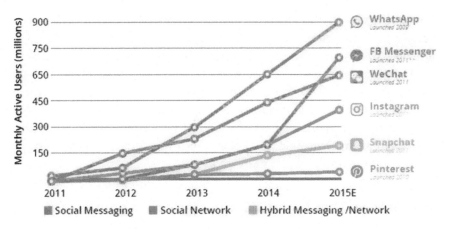

Fig. 4.2 Communication explosion over time (Van Doorn 2016)

to find the right app. Often, the extended benefit of an app—in addition to the company website—is not clear (Fig. 4.2).

Every newly installed app also means having to get used to a new user interface. Messaging apps in contrast are all similar in the set-up and layout and the operation of them is simple, even for new users.

4.2.3 Subject-Matter and Areas

Conversational Commerce describes a new trend in the area of consumption. The term was coined by Chris Messina who is currently a developer at Uber and gained distribution and acceptance by way of the hashtag #ConvComm (Messina 2016b). In essence, the concept is not a new one as every form of trading traditionally starts with a conversation. In the times of online shopping, conversation has taken a back seat as the large numbers of customers cannot be taken care of in one-to-one conversations and in real time. When purchasing on the Internet, one-sided communication is increasingly fallen back on where the customer fills in contact forms or sends e-mails. Direct contact with companies by phone is often possible but is frequently associated with charges and long waiting times on hold. All in all, the forms of contact dominating nowadays are connected with waiting ties for the customer and thus of a disadvantage in comparison with classical sales talks.

Conversational Commerce, in contrast, offers individual, bidirectional real-time communication with the customer without there being a need for unrealistically high numbers of employees. The conversation can take place with the help of chatbots that are either integrated into platforms such as WhatsApp or Facebook Messenger or can be found on their own on the company's website. In the chat conversations, product advice, the saes process, purchasing process and customer support can take place and thus alleviate consumption for the customer. As the customer interacts with the company or the brand in the same way as with a friend, we speak of the "brand-as-a-friend" concept (van Doorn and Duivestein 2016). Companies thus benefit from their chatbots being able to lead conversations that feel natural and humanlike for the users.

4.2.4 Trends That Benefit Conversational Commerce

Conversational Commerce is mainly pushed by large Internet companies that operate a messenger and/or chatbots such as Facebook, WhatsApp, Telegram, Slack, Apple and Microsoft. The headway in Conversational Commerce is led in the main part by two developments: a communication trend and the boom in artificial intelligence.

The former can be recognised in the popularity of messaging services, the use of which is increasing at a virtually explosive rate. Apps and services that serve the communication with friends and acquaintances have established themselves, in contrast to most other apps. As the proportion of mobile natives (users that grew up with mobile digital services) is increasing, the use of messaging services will probably continue to increase. Due to the large number of people that use messaging apps, the next logical step for companies is to offer their services there. Instead of convincing the customers to install a new app, the companies pick up their customers where they are already to be found, as chatting is already integrated into daily life.

Development in the field of AI also makes the existence and further development of Conversational Commerce possible with regard to the performance of speech recording, for example, that increases by 20% every year. It is already possible nowadays to capture more than 90% of spoken and written language thanks to the processing of natural language, also called NLP.

Aside from the two essential criteria explained for the growth of commercial commerce, there are further trends that benefit its progress. One example is the so-called quantified self movement that records and analyses personal data throughout the day, data such as food consumed, air quality, moods, blood oxygen levels as well the mental and physical performance. In some cases, wearables, i.e. devices worn on the body enable the recording of these levels by way of, for example, electronics and sensors worked into the material of the clothing. Together with the progress in the field of data science, this trend has the potential to personalise customer interactions in Conversational Commerce as well as to predict the consumer's needs.

Of essence for the implementation of entire purchasing processes in the framework of commercial commerce is the integration of seamless payment technologies. These are available for third-party providers to an ever-growing extent by way of APIs.

4.2.5 Examples of Conversational Commerce

The probably oldest implementation of Conversational Commerce took place via WeChat, a mobile cross-platform messaging service from china that was brought to life by the holding Tencent in 2011. Via WeChat, friends and acquaintances can be communicated with as well as service from countless companies can be used. You can, among others, call a taxi, order food, purchase cinema tickets, make doctor's appointments, pay invoices and record your daily exercise program. WeChat is a chat-based interface with many additional features such as mobile payments, chat-based transactions, media and interactive widgets.

By way of a powerful API, it is possible for the diverse companies to "become friends" with their customers. More than ten million companies have joined the chat platform and the popularity among small businesses is increasing. In contrast to the USA and Europe where to date, services are mainly offered in specific apps, the merging of messaging and consumption was focused on much earlier in china. In the meantime, WeChat has become one of the largest standalone messaging apps in terms of the number of active users. In the second quarter of 2016, 806 million active users were registered (China Internet Watch 2016). Instead of changing existing infrastructures like in the USA and Europe, in China markets can be initially entered by way of mobile applications and payment systems, according to Brian Buchwald, CEO of the consumer intelligence company Bomoda (Quoc 2016).

Facebook Messenger recently got competition from Google Allo, a "smart messaging app" according to the manufacturer, that integrates the Google assistant. Chatting with friends can thus be made simple by the likes of answer options provided by the bot that can be selected by the users by clicking on them. The Google Assistant can be brought into the conversation by addressing @google to find videos, for example, to obtain directions or retrieve information. A direct conversation with the Google Assistant can also be started up and help for various queries obtained.

Echo, the personal assistant of Amazon, is also an example of Conversational Commerce. Aside from the assistance at home like the playing of music or requesting recipe ingredients, the device can also be used to access the entire Amazon catalogue of goods and to purchase goods. This way, frequently used goods can be reordered via a conversation with the built-in bot, Alexa, in a simple manner. Furthermore, Echo is connected to the services of other companies via the development platform Alexa Skills. This enables the requesting of an account balance, and the ordering of an evening meal using a simple commando.

Other platforms that enable customer interaction in real time via bots for a wide range of companies are Operator, Slack, Snapchat Discover and Snapcash, AppleTV and Siri, Magic, AIk Bots and Telegram (Quoc 2016).

4.2.6 Challenges for Conversational Commerce

All chatbots function in a similar way. They are based on the comparison of patterns in the text and react towards certain keywords. Yet, what challenges are the active chatbots facing and why is Conversational Commerce still not more common?

One reason seems to be that the integration of AI has not yet been widely realised. The author of an article in the magazine c't, for example, criticises that at present there is still no bot that can learn the interests and preferences of users and can operate proactively without being triggered by the user (Bager 2016). In an article in the magazine "Absatzwirtschaft", the author describes that the integration of AI in bots is lagging behind (Strauß 2016).

By observing decisions and activities, bots can get to know the user better. Another challenge is seen by the author in the ability of the bot to adapt; the program should be able to adapt its own settings to external influences. Another demand on bots is that they act foresightedly and start-up processes at their own initiative, such as reminded the user to buy coffee. The bots are also to become social so that they can develop a kind of "social life" among themselves and communicate with each other. It is, however, questionable, whether these are the reasons for Conversational Commerce not being more widespread, not least in Germany. From a technical point of view, the chatbots' ability to learn, adapt and be foresighted is by all means feasible.

There are thus a large number of libraries for developers to integrate the ability of chatbots to learn and be foresighted.

4.2.7 Advantages and Disadvantages of Conversational Commerce

It goes without saying that the use of chatbots in Conversational Commerce brigs along not only many advantages for consumers but also for companies. The humanlike conversations, the better and faster service as well as the presence of the brand can lead to closer customer retention. Many consumers appreciate the services tailor-made to them. The improved services lead after all to an increase in customer satisfaction. The reputation and the profile of the brand or of the company can also be increased. And moreover, companies obtain more insights into their customers' wishes and needs as well as into the purchasing process and context.

However, it must not be forgotten that Conversational Commerce can also entail disadvantages or potential problems. One example is the concerns of the consumers with regard to data protection and privacy. Transferring chat trails to companies is inconsistent with German law. And the probability of data misuse could increase as criminals could gain access to payment details and other information. To date, it is not clear how transparently the activity of robots in Conversational Commerce should be dealt with. Should the consumers be informed that they are currently chatting to a bot? As customer care by telephone will lose significance thanks to the use of chatbot, a loss of jobs can also be expected. For companies it is thus important to develop strategies to prevent frustration among the staff, for example, caused by finding new jobs within the company.

4.3 Conversational Office

4.3.1 Potential Approaches and Benefits

Bots not only provide help in personal organisation such as with personal butlers) or in marketing (e.g. via algorithmic marketing) and with sales (such as in Conversational Commerce) but are also perfectly suitable for use within a company. For this area of use, Amir Shevat (2016) coined the term 'Conversational Commerce'. Shevat divides the digital developments in companies into three different eras: After the computerised era came the mobile office era, which is now making room for the conversational office era. As modern communication in offices is mainly text-based, systems that enable simple messaging to individual colleagues or groups can help to save time.

An example of such a collaboration and organisation system for the office is the Slack software. The platform was established in 2013 and has about one million users every day. What could create the breakthrough for the concept of conversational office is that, irrespective of the conversation between employees, bots can also be involved in the conversations. At present, there are various types of bots that can be of assistance with organisational tasks. The business trip bot Concur by SAP, for example, can plan trips for employees, the expenses bot Birdly by Slack can process submitted travel expenses and the human resources bot Ivy by Intel can help staff with various questions, e.g. with regard to the salary.

Office work, in particular when several people are involved, can function more efficiently if the bot is involved in the conversation without being asked and provides assistance. This can also reduce the frustration that many employees experience when they have to interact with user-unfriendly software in lengthy processes to apply for their holiday or submit invoices. If all staff members are actively present in one system, new employees can be integrated more easily and their expertise and opinions better incorporated.

One scenario of conversational office is, for example, if colleagues discuss via an online system an error in the system they are working with. A bot could provide all details about the error without being asked as well as record when the error was rectified. The notion of digital employees and bosses will be explained in the following.

4.3.2 Digital Colleagues

Ben Brown (2015), co-founder of the software company XOXCO, implemented a mixture of messaging, automated software and artificial intelligence in the shape of an intelligent digital employee that he christened Howdy. He can be integrated in the Slack platform and is meant to take over the most boring, repetitive and mundane tasks such as planning a meeting with several attendees or collecting status reports. Howdy can communicate with all attendees simultaneously to find a data or collect information about tasks done and problems that have arisen. The fact that the concept of the digital colleague will assert itself is suggested by the multitude of office bots such as Weld, Geekbot, Flock, Tatsu Nikabot, awesome.ai, phonebot, ElRobot and Pushpop, to mention merely a small choice (Vouillon 2015).

According to IBM, the company has created the "ideal employee": Celia (Cognitive Environments Laboratory Intelligent Assistant) has a store of knowledge that is based on the advanced analysis of millions of pages of text. This means that she can give doctors medical advice and suggest new taste variations to cooks. In comparison with her predecessor Watson, who defeated two top-class candidates in the quiz show "Jeopardy", Celia can have better dialogues and explain her answers. This makes her more humanlike.

In China, a bot has been used since 2015 to present the morning weather. Xiaoice presents the weather based on official meteorological sources and provides further suggestions such as avoid outdoor activities if the air quality is bad.

It is not unthinkable for bots to take over the role of the boss A board member of the risk capital investor Deep Knowledge Ventures from Hong Kong comprises, for example, a program called VITAL. The abbreviation stands for Validating Investment Tool for Advancing Life Sciences. The software can provide the human board members with the necessary information to make better investment decisions. Dmitry Kaminsky, a senior boss in the company, sees the combination of machine logic and human intuition as the ideal combination. In Japan, an advertising agency has in the meantime employed a robot as Kreativchef (van Doorn and Duivestein 2016).

The new concepts of bots as employees, bosses or chairmen of the board are still not mature and raise many questions. Van Doorn and Duivestein from SogetiLabs are bringing to attention that it is still unclear who assumes the responsibility if bots make mistakes. Can they be controlled when they develop their own social life and communicate with other bots and possibly make wrong decisions? The many advantages of bots as a helping hand in office work are not to be dismissed and it could be as equally dangerous to do without the office bots.

4.4 Conversational Home

Intelligent bot systems, however, are not only used in companies but increasingly also by consumers. Amazon Alexa or Google Home help consumers, for example, as digital assistants to simplify looking for information or ordering products (both systems have been available in Germany since 2017).

4.4.1 The Butler Economy—Convenience Beats Branding

Traditionally, by the word butler, we understand a personal servant who is available all of the time and fulfils our wishes. A conscientious butler knows us so well that he can even foresee needs and make recommendations. With bots that are adaptive and that can thus be described as intelligent, the step towards the personal butler, the digital personal assistant is no longer far away. The large technology companies Amazon and Google have had digital butlers for home use on the market since 2016: Echo and Home are standalone devices that remind of loudspeakers and which regulate the lights, temperature and music as well as weather queries, alarm function and requests for information. In addition, Google Home can send e-mails and text messages as well as sort out photographs and use card services.

Examples of personal assistants that can be integrated into a telephone or computer are Siri (Apple), Now and Allo (Google) as well as Cortana (Windows). Siri, a digital assistant that does have a sense of humour but has difficulties with speech recognition, will be replaced by his big sister Viv in the near future. Dag AIttlaus, managing director of Viv, announced during the official demonstration of the personal assistant in May that Viv "bring the dead smartphone back to life by way of conversation". The name of the algorithm is the Latin root of the word life. Even Facebook is currently experimenting with their own personal assistant called "M" and which will soon be available worldwide.

A personal butler, also called personal assistant or digital servant, is a program that is integrated in a technical device, an operating system or an app, and which can take over daily tasks such as shopping, bookings, bank transactions, planning or regulating light and temperature. With time, the personal butler gets more familiar with his owners and can predict their wishes and needs.

What is the same with all virtual assistants is that they are meant to take over everyday tasks such as booking hotels or taxis, ordering clothes, food or flowers, or even bank transactions or writing to-do lists. Instead of comparing offers for hours on end, entering account details or finding the right app for notes, these frequently cumbersome tasks can be done in the time it takes to say a sentence. And when humans are no longer needed for these tasks, the human resources could put to use in a different way, for creative tasks for example.

The following section will first explain why the comfort of the personal butler will diminish the significance of the brand. Afterwards, the past and present development process of digital butlers will be described and whether verbal or written communication with the assistant will dominate in the long term will be discussed. At the end of the section, the advantages and disadvantages that could result for consumers will be illustrated.

4.4.1.1 Comfort Is Becoming More Important Than the Brand

All of the large technology companies are currently competing for the best personal assistant. The field is lucrative as people with a personal assistant will spend even more time on their mobile and thus both income from advertising and device sales can increase. The search for keyword terms, via Google for example, will presumably disappear in the long run in the scope of this development. Instead of that, purchasing decisions will be made in conversation with the digital assistant. Product recommendations in social networks may also lose significance. It is probable that products suggested by the persona assistant are a better match for the users than ever before, as the persona assistant avails of a larger amount of information than what personalised advertising is based on. If more products are presented to the consumers that are much more tailor-made to his or her needs, it is probable that in sum, more products will be consumed.

For a brand or a company to be successful in the future, it is thus important that the respective products and services are taken into consideration by the personal butler's algorithm. If a user then wants to order flowers, book a hotel or buy a coat, the personal assistant will only consider those companies that are present in the network of the algorithm. For personal assistants of Google, on the other hand, the ranking of the results in the Google search can play an important role. In the future, the focus of customers will be less on the brand than on convenience. This means that companies that understand how to be connected to the relevant personal assistants will win.

Amazon, for example, could soon offer own labels via comfortable ordering processes without having to surrender margins. The first step in this direction is the Amazon Dash Button, a button that is placed on devices to order goods to be refilled at the press of a button such as washing powder or toilet paper, which was introduced in 2016. The team behind Viv[1] is still trying out various business models, but one could involve a processing fee for every enquiry.

4.4.2 Development of the Personal Assistant

The two crucial requirements that enable the existence f the digital servant are, on the one hand, the linking of different services to a huge network and, on the other hand, the learning aptitude of the assistant. For a digital assistant to be able to answer enquiries, it is essential for various programs, apps and other services to be able to communicate with each other. In order to be able to book a taxi with Apple's Siri, for example, the operating system must allow access to services such as Uber, which was ultimately realised in iOS10.

In the official demonstration of Viv, Dag AItlaus provides an insight into the huge network of categories and subcategories for different services and information that is behind the future personal assistant. With an adaptive assistant, even needs can be predicted after some time. The digital butler is thus personalised so that products can be suggested, for example, that are a perfect match to the user's needs.

The development process of personal assistants has been divided by research institutes and observers into different yet similar categories. The research institute Gartner, for example, calls the development from the simple smartphone to the perfectly personalised butler as cognisant computing (Gartner 2013). They have subdivided the process into the four steps Sync Me, See Me, Know Me and Be Me. Sync Me implies that copies of all relevant content is stored in one place and can be synchronised with all used end devices. This has already been realised in the course of cloud computing, to be more precise, since it has been possible to store backup copies of telephone and computer data in so-called clouds. The second step See Me assumes that the algorithm knows where we are and where we were in the past, both on the Internet and in the real world. This is also integrated in the use of smartphones and computers to a large extent. The third step Know Me is currently being implemented with the first personal assistants as well as with services such as Netflix and Spotify, which are meant to understand what the user wants and to suggest matching products and services (films and music in this case) accordingly. Be Me is currently a scenario of the future in the main part, in which the butler acts on behalf of the user according to both learned and explicit rules. If the assistant independently improves itself, the answer and recommendation mechanism can be finetuned further. Amazon's Alexa, for example, can get to know the user's needs better and better and tries to adapt itself to them. Via the Alexa Skills developer platform, the personal assistant can also learn a new task as well as be connected to other companies.

Development of the personal butler in the scope of cognisant computing:

1. Sync Me: Backup copies are stored in the cloud.
2. See Me: The butler observes the user's activities, both on the Internet and in the real world.
3. Know Me: The butler suggests suitable products and services.
4. Be Me: The butler acts independently on the user's behalf according to explicit and learned rules.

At present, personal assistants are still acting passively. This means that do not become active until apps are accessed, certain buttons are pressed or they are greeted. Active personal assistants with artificial intelligence could also join in conversations on their own and give advice or clarify misunderstandings. However, this also entails risks: The assistant could make inconsiderate statements if, for example, the user gives another person evasive answers or uses white lies and the personal assistant interrupts the conversation and exposes the user.

For the perfect integration of personal assistants, it will be essential that the butler is omnipresent, i.e. is synchronised on all devices. If you forget your smartphone at home, another device such as the smartwatch should be equipped with all information and skills. Even gestures are to be perceived and understood by personal assistants in the future, with the help of a camera and sensors. Another desirable function is speech recognition to protect the access to private functions such as the diary for example.

4.4.2.1 Speech or Text?

One question that is controversial among the observers of the development of the personal assistant is the type of communication. Will speech or writing dominate? The computer geek Graydon Hoare states among the advantages of text in comparison with speech that it is possible to communicate with several parties, that text can be indexed, searched and translated, as well as that text allows highlighting and notes and that summaries and corrections can be made (Hoare 2014). Likewise, Jonathan Libov (2015), who works as a risk capital investor for Union Square, prefers text over speech. He points out that the comfort in writing is more important than the convenience of speaking ("comfort, not convenience"). He sees text-based communication as more comfortable as it saves time and is fun speaking in contrast, does not require as much effort and is thus to be regarded as more convenient. The text-based communication, on the other hand, is also flexible and personal. According to Libov, NLP is not good enough for us to be able to rely on oral communication with technical equipment. Instead of that, innovations in text-based communication enable faster answers such as QuickType, a program in Apple's iOS operating system that can extract the options placed in a message so that the user does not have to write the reply themselves but merely has to select it.

Supporters of verbal communication emphasise that speech can be more natural and faster. Especially for applications within a home, for example for regulating lights or music, spoken commands seem to be more natural and easier, according to van Doorn and Duivestein (2016) from SogetiLabs. Speech recognition will work increasingly accurately and functions in some devices from a distance, as well, such as on Amazon's Echo. In fact, the four major personal assistants of the present day are speech-based: Siri, Now/ Home, Cortana and Echo.

The bot enthusiast Chris Messina points out that when driving, one cannot give instructions per text and does not want to record any notes via a microphone in a presentation. In the end it seems as though there is need for both text- and speech-based communication.

4.4.2.2 Advantages and Disadvantages for the Users

According to Chris Messina, the two greatest advantages of the personal butler for the users are convenience and adaptability. Due to the fact that with a personal digital assistant, apps no longer need to be searched for, downloaded, installed and configure, the time between a question and the answer can be reduced, increasing the convenience for the user. The adaptability of the personal butler increases if the butler is increasingly personalised and an awareness of correlations is developed. Messina specifies that the user has to adapt to the app when using apps. Instead of that, it can be expected of a personal butler that it adapts to the user as we are used to in inter-human interaction. Our friends, for example, would not bombard us with text messages if they knew we were currently driving a car, but would wait until we were available. According to Messina, it is essential for users to be able to state when they do not wish any information from the digital assistant or that they expect the information when various framework conditions have been fulfilled. Such framework conditions could be, for example, that the user has arrived at home, which could be automatically determined by the butler using GPS.

Another important point according to Messina is that the PA can adapt to the user's mood and the current context. The user can, for example, be tired or out for an evening meal with friends and thus possible not be interested in going through every option with the personal assistant. Instead of that, the algorithm could automatically make decisions at its discretion without constantly getting back to the user. If the algorithm is aware of these circumstances, its reactions can also feel more empathetic and accessible. In other words, the technical devices that we use should adapt to our circumstances the way other people do.

A potential problem on correlation with the emergence of digital personal assistants is the filtering of content that may restrict access to freely available information. Should Facebook become the new Internet, the question is in whose interest will it act? For some people, the lack of privacy when using personal butlers could pose a problem. After all, the personal assistants see and know everything about the user and the data is not only evaluated each piece on its own but is also linked up, which can provide even deeper insights into the user's personality and life. On the other hand consumers tend to give up lot for their convenience. This is why what is offered to the user must be worthwhile, in order to consent to data disclosure. Carolina Milanesis, vice president for research at Gartner, is of the opinion that the data available about us that is used by our devices, "the likes and dislikes of our environment and relationships", will improve our life in the end (Gartner 2013).

4.4.2.3 Siri, Google Now, Cortana, Alexa, Home—Who Is the Cleverest of Them All?

The personal assistant and digital butlers described are offered by the well-established technology companies Amazon, Apple, Google and Microsoft. Apple and Google have on offer the digital butler Siri and Now respectively for iPhones and Android phones, and Microsoft has developed the assistant Cortana for the home operating system Windows. Meanwhile, the online retailer Amazon has concentrated on device for the home, the loudspeaker Echo with the built-in digital assistant Alexa. Yet, amazon does not remain without competition in the home area, for Google has already launched a similar product called Home in the USA. hat in den USA. Every company is claiming to have the best digital assistant in their range. Yet, how helpful can the butlers of today with enquiries of any kind? In order to get to the bottom of this question, the Institute for Digital Business at the HTW Aalen tested the digital assistants Siri, Now, Cortana and Alexa that are available in Germany in various categories of questions.

4.4.2.4 Procedure and Set-up of the Study

In order to find out which of the most common personal assistants is the cleverest, we set up the systems Siri, Now, Cortana and Alexa and used them in everyday routine for two weeks. In the meantime, we identified five different categories of questions to test the assistants: "Classic", "General", "Knowledge", "Commerce" and "Untypical". We these types of questions, various functions of the assistant could be examined, as resented in Table 4.1.

Besides general assistance, the functions of the assistant were also tested as a friend for recommendations, as a lexicon for knowledge questions or for purchasing assistance. Furthermore, untypical questions were asked that were meant to test the intelligence of the digital butlers.

Five to twelve questions were defied for every question category, which varied in the degree of specialisation. At the same time, it was presumed that the more specialised the question, the lower the probability was that the assistant would process the question correctly or give the right answer. The structure of the questions according to their degree of specialisation is presented by way of example in Table 4.2 on the basis of the questions asked in the "Knowledge" category.

The respective degree of specialisation of the questions was adapted, among others, with the help of the identification of the frequency of the words used in Duden. This way, a high degree of specialisation was allocated to a question that contains a word of low frequency in Duden. An example

Table 4.1 Question categories for testing the various functions of the personal assistants

Question	Classis	General	Knowledge	Commerce	Untypical
Function	General assistance	Friend (recommendations)	Lexicon	Purchasing assistance	Intelligence

Table 4.2 Questions from the "Knowledge" category with increasing degree of specialisation

Degree of specialisation	Questions from the category "Knowledge"
Low	How many inhabitants does Stuttgart have?
	How many inhabitants does Teheran have?
	How big is Germany?
	How big is Andorra?
	How long did World War I last?
	When was the fall of the Berlin wall?
	Who is the Home Secretary of Germany?
	Who is Otto von Bismarck?
	What does laicism mean?
	What does perception mean?
	What is the EU Commission?
High	What is TTIP?

of this are questions about the size of Germany and Andorra. As the word "Andorra" is a lower-frequency word than "Germany" in Duden, the question about the size of Andorra is allocated a higher degree of specialisation. Complex questions were also allocated a higher specialisation, the answers to which requiring additional steps. One example to the question about how long World War I lasted. To answer this, the artificial intelligence system must first find out when the war started and ended and then work out the time it lasted.

In order to obtain reproducible results, the questions were asked on the digital assistant more than once. The assistant's answers were given a score from two to zero points. Two points were given if the answer was good, i.e. when the answer to the question was appropriate and the assistant proved to be of assistance. One point was given is the assistant did indeed understand the question but could not or could only partially be of assistance. An answer was given zero points if the digital assistant was not able to help at all of gave fully meaningless answer to the question.

4.4.2.5 Results of the Study

In order to determine which of the digital assistants is the best, we compared both the overall results of all question categories and the results in the individual question categories with each other. In addition, we compared the performance of all assistants in the entirety in the various question categories to find out which question categories are best mastered by the digital assistants. For a fair comparison, the points score in each question category were divided by the number of questions to obtain an average score (Fig. 4.4).

If we summarise the performance in all tested question categories, Amazon's Alexa clearly comes out ahead. The assistant is closely followed by Google Now and Apple's Siri which are almost tied, and at some distance from Microsoft's Cortana in last place. Alexa's joy of shopping, Now's vast range of knowledge, Siri's versatility and Cortana's reserved intelligence become clear in a final comparison of the average scores in the different question categories (Fig. 4.3).

Alexa and Siri scored best in the category "General", which contained questions like "How are you today?" or "What do I have to do today?" or "What birthday present can you recommend to me for my wife?". Siri was in fact not very helpful with personalised questions, but gave accurate answers to more general questions. Alexa only did not know what to do with the question about a birthday present and responded almost exclusively

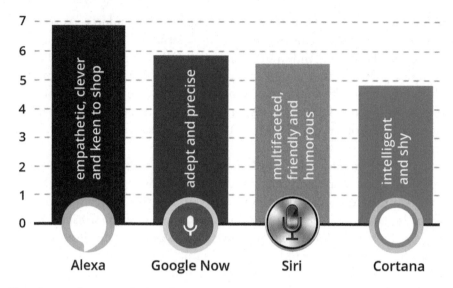

Fig. 4.3 Total score of the digital assistants including summary in comparison (Gentsch)

appropriately and accurately otherwise. Cortana and Now were frequently not able to answer questions in this category but, in some cases, forwarded to corresponding applications or search engines.

The general help questions in the category "Classical" that range from "Will it rain tomorrow?" over "What is X times Y?" to "My mobile is broken, can you help me?", were best answered by Siri. She presented solutions to all questions, but sometimes she only answered the questions in part. This was the same for the slightly weaker assistants in this category Alexa and Now, whilst Cortana did not answer most questions are gave wrong answers.

The category "Commerce" which, among others, included the requests "Order me a stethoscope!", "What shops are close by?" as well as "What does an iPhone 6S cost?" is clearly dominated by Alexa. The digital assistants reacted to all questions and requests accurately and only had difficulties in finding shops nearby. Cortana and Now were tied in mid-field in the commercial category with consistently food reactions, yet with slightly different weak points. Whilst Cortana did not understand the word "stethoscope", Now just like Alexa gave no meaningful answer to the question about shops. Siri, in contrast, was an expert for this type of question, but according to own wording, ordering products was beyond her skills and no assistance was provided in this respect.

Comparison of the assistants according to question category

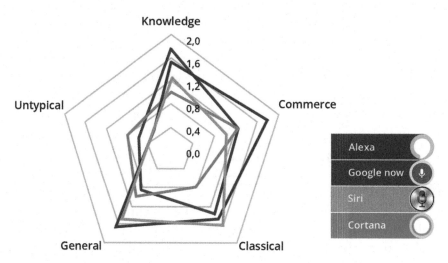

Fig. 4.4 The strengths of the assistants in the various question categories (Gentsch)

The category "Knowledge" with questions like "How many inhabitants does Teheran have?", "Who is the Home Secretary of Germany?" and "What is TTIP?" was led by Now. The Google service was given full points for almost all questions and only had difficulties with the current Home Secretary and the abbreviation TTIP. Alexa came in second place in this category, but was not able to answer any questions to do with data or periods and, just like Now, did not understand the abbreviation TTIP. The acronym was, in contrast, understood by Siri, but the assistant merely referred to a non-associated page on Wikipedia (Fig. 4.4). For most of the other questions, Siri referred, however, to appropriate Wikipedia entries, yet rarely provided the answers in speech. Apart from one exception, Cortana was able to answer all questions. However, Cortana was only able to give the answers that were found via the Bing search engine in writing and not verbally.

Questions that challenged the artificial intelligence of the assistants are contained in the category "untypical" and are for example, "Can you recommend a new laptop to me?" or "Do I have a free day in my calendar?". For Cortana, this is the only category where the Microsoft leads the board. More than half of the questions were responded to, albeit not exhaustively, yet at least meaningfully with search requests via Bing. Now understood many questions at least to some extent and provided solutions in the shape of websites. Siri and Alexa, the bottom of this category, were hardly able to help in a meaningful way.

In order to ascertain which of the question categories the digital assistants mastered best, a quantitative comparison of the overall score of the assistants was drawn in the various question categories. It shows that the bots render the best performance on average in the category "knowledge". The categories "Commerce", "Classical" and "General" are however not far behind and follow almost at a tie. In comparison with the bet performance in the field of knowledge, the bots, in contrast, achieved less than half of the points in the category "Untypical".

4.4.2.6 Conclusions and Outlook

The results of our study that was meant to find the cleverest among the digital personal assistants show that personal digital assistants tested have particular fields of speciality and weaknesses in different areas. The cleverest assistant respectively in the different categories are shown in Fig. 4.5, and in the following text, the strengths and weaknesses of all assistants tested will be presented and discussed in more detail.

All in all, it has been shown that Amazon's assistant for at home, Alexa, is the clear jack-of-all-trades and winner among the personal assistants tested. In most areas, i.e. with classical help requests, recommendations,

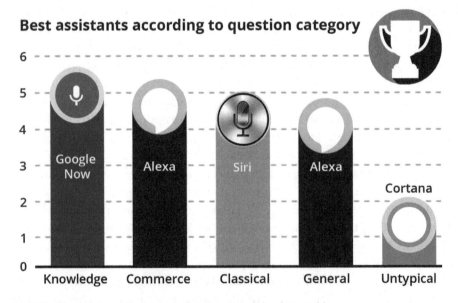

Fig. 4.5 The best assistants according to categories (Gentsch)

social conversation and when asking for facts as well as a shopping assistant, Alexa gives accurate answers or reacts in the way expected. The assistant only stumbles a little when it comes to more complex questions. The reasons for Alexa's high performance probably lie in the increasing number of third-party developers that program the applications—so-called skills—for the assistant and thus make it ever cleverer. At the end of February 2017, around 1000 different skills could be found on the German side of the company, which also facilitate the integration of the assistant with external provides such as Bild, Chefkoch or BMW. It is thus not surprising that most of the questions asked, especially in the field of commerce, could be answered by Alexa immaculately. It is to be expected that Alexa will continue to be improved in the future by the integration of third-party providers. More problematic is the distribution of the hardware as Alexa cannot be installed on smartphones but comes in the shape of the Echo loudspeaker. Until now, the sales figure for echo have been comparably good and it is not improbable that the device will dominate the market as the "operating system for the smart home" (iBusiness 2017). Echo's built-in smart home control unit as well as the high presence and increasing number of Alexa's skills that are uploaded from home device manufacturers in the category "smart home" all peak in favour of this.

Google Now, the personal assistant for Android smartphones, was able to come in second place in our study particularly due to its brilliance in knowledge questions. There were some minus points for lacking personalisation, i.e. the fact that individualised recommendations could often not be given. In addition, the answers were often inaccurate. The assistant cannot independently process purchasing commands either. As Now is directly connected to Google, the largest search engine, the good result in the field of knowledge is not surprising. Apart from the head start in data, Google is also home to services such as YouTube, Google Maps and PlayMusic. As the application can easily be integrated into the assistant, it is too be expected that Google will continue to catch up in the race. Another advantage in comparison with Alexa is that Google software can be used on many different types of hardware as well as laptop, smartphone, TV, etc.

Siri the competitor in Apple smartphones is, in comparison with the assistant for Android, rather an all-rounder and ended up just behind Now. Siri is especially distinguished by a friendly and humoros manner and was able to handle classical help requests without difficulty. However, she also had difficulties with individual questions such as with recommendations. She did provide assistance with knowledge requests and shopping commands, but was often not able to execute the overall process independently. Siri will be replaced soon by viv.ai, which according to the company bearing the same name, will belong to a new and cleverer generation.

The digital butler Cortana by Microsoft, which assists in Windows devices, was frequently unable to answer verbally and referred to websites, frequently via the search engine Bing instead. The assistant was also unable to process personalised request. For this reason, it ends up in last place in our study, although the bot appears to be relatively intelligent and was able to handle some search requests that went beyond the skills of Alexa, Siri and Now.

4.5   Conversational Commerce and AI in the GAFA Platform Economy

The aim of the GAFA economy (Google, Amazon, Facebook, Apple) is to know the ecosystem of the consumers as well as possible and to also be able to operate it accordingly. Whoever can master this task best can also place their own products best with the consumer. It is for good reason that the GAFA world is developing systems to monopolise access to consumers. This new form of market capitalisation is accompanied by the risk of misuse of market power and can result in high penalties as Google recently able to feel the effects of.

Anyone who has a direct interface to the customer in the shape of a bot or messaging system, which knows consumer preferences and behaviour across all fields of life, determines the information, advertising and purchases. If the consumer selects their favourite from the list of hits themselves during a Google search or an Amazon product search, the bot recommendation is usually reduced to one product or one piece of information. The bot sovereignty thus replaces the active evaluation by the consumer. The fact that this battle is highly relevant and lucrative is shown, for example, by the efforts of Amazon to win control over the customers by way of the dash button and the DRS system under the disguise of convenience. This shows how Amazon is trying to penetrate the consumers' ecosystem. The manual automation of ordering new washing powder at the press of a button is only the beginning. In the next step, there is a speech-controlled dash button available. Yet the system can do much more: An automatically acting DRS system (Dash Replenishment Service (DRS)) enables connected devices to order products from Amazon (if they are running low), recognises the need for products, i.e. it knows the stock of, for example, washing powder, toothpaste or printer cartridges. If a product is running low, the order process is triggered (Fig. 4.6).

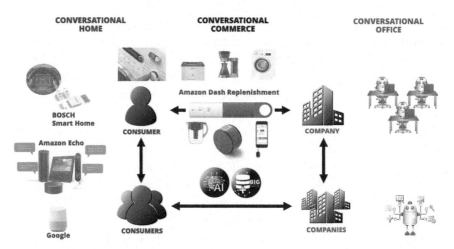

Fig. 4.6 AI, big data and bot-based platform of Amazon

One of the greatest strengths, but also the greatest point of criticism, of the Alexa ecosystem is the integrated and automatic AI-based analysis of customer interaction. The digital data track of the customer, for example, can be used so that their Alexa really gets to know them. This way, the cloud not only stores the settings of the DASH buttons but also derives preferences and needs the customer has from purchasing behaviour and search queries. With the help of AI, high-quality forecasts about further customer communication can be made from this information and this information can be incorporated in cross-selling strategies.

Likewise, location-related data and services can be collected and offered through positioning services. The possible number of data points to be recorded that can be correlated with customer behaviour seems to be almost endless thanks to the strongly distributed user experience in the ecosystem.

Yet, it is not only the text- or data-based analysis of customer behaviour that is relevant. Due to the massive progress in NLP, not only can the factual level of the customer statement by analysed but also the customer's current mood can be determined. This provides an emotionalisation of the bot-customer relationship by way of the bot training empathetic behaviour that comes closer to interpersonal communication.

With the deep interlocking in the customer's ecosystem, the unique possibilities of data acquisition and analysis result for the companies. Due to the centralisation and monopolisation of the customer interface, companies can hold the consumer in their "consumer bubble" on the basis of extensive preference and behaviour profiles and capitalise them.

One consequence of this development could be that the emotional brand commitment loses in relevance, resulting in an objectification of marketing. For, purchase decisions are now made more rationally than they were before. Due to the development of smart homes or smart products, there are more rational purchase decisions—bots are now representing humans increasingly more. The refrigerator "decides" when more milk has to be bought. A digital representative of customers is logically immune against emotional and empathetic advertising, which loses its meaning due to that. The ideal value of the brand is irrelevant for the customer bot which, in the ideal case, thanks to the customer's digital signature, acts objectively as their representative in e-commerce. This way, the access of companies and companies to the platform becomes more important than the brand itself.

Data-based marketing (intent-based marketing) is on the rise. Marketing departments are already collecting masses of behaviour-based data. When Alexa, Siri and Google Assistant find their way into the living room, the comparison to a Trojan horse is not far off. If providers notice, for example, that there was a marriage, offspring is also possibly in the offing. This information can be worth its weight in gold. It remains to be seen how preferences due to more convenience can be conciliated with the risk of market misuse of monopoly-like commerce ecosystems. The fact that consumers are open to new convenience technologies is shown by the trend towards voice-based interactions. This year, every fifth enquiry via Google was via voice. A 50 percentage quota is forecast for 2020. In ten years' time, it will probably be around 75 percentage of all Google searches.

Whilst present-day communication is till between the consumer and the company bot, in the years to come, there will be increased communication between the consumer bot and the company bot. for this reason, marketing activities must be adapted to bot channels. A process of rethinking will also have to take place with SEO or SEM. The so-called bot engine optimisation, BEI in short, transforms the guiding principle "rule the first page on google" into "rule the first bot answer". The focus lies on personalised one-to-one campaigns from bot to customer.

Of course, companies have always analysed data about consumers in the scope of database marketing and analytical CRM data, in order to align products and communication with target groups and to thus be a profitable as possible. Only, companies and consumers are no longer meeting each other on classical markets but the providers is internalising the market in a certain way. Amazon has not been a retailer of products for a long time now, but a smart ecosystem that intelligently captures, analyses and uses data to keep the consumer in their own commerce bubble.

4.6   Bots in the Scope of the CRM Systems of Companies

When bots are increasingly used in companies, the CRM system will also increasingly turn into a "BRM – Bot Relationship Management" system. With each contact with the customer, the bot learns more about the customer's needs and preferences. It acts as a fully automated, smart customer adviser who can recognise the client's wishes like a good friend and fulfil them directly. Fully personalised up- and cross-selling increase customer satisfaction and frequency of purchase. With the help of these persona assistants, the CRM system of a company is given fully autonomous efficiency never achieved before and alignment as close as possible to the customer.

The search for a suitable and affordable flight can be cumbersome. What is we can simply ask a bot for an affordable flight? Lufthansa with their helpful avatar "Mildred" (mildred.lh.com) have recognised the signs of the times and gone public with a best-price search bot at the end of 2016, initially a still learning beta version. In a sympathetic chat with Mildred, you can enquire in German or English about affordable flights within the next 12 months and book them directly.

Admittedly, the speech requirements are not particularly high as the chat does not take any unsurprising turns. Of course, the search period can be limited further and the booking class can be specified but the content of the chat is more or less the same. It is connected to various databases including "Lufthansa Nearest Neighbour" to search for airports according to city names or the three-letter cods. With the help of "Google's Geolocating", Mildred is able to located airports according the sights. An enquiry about the Eiffel Tower, for example is translated into Paris as the flight destination.

On the basis of this data, Mildred enquires with the Lufthansa database "best Price" about the cheapest price for the route needed, which can then be booked via a link.

The classical inbound touchpoint bot in customer service is provided by the service provider for digital television, Freenet TV. It gives advice about reception problems around the clock and can thus provide initial help. In contrast to Mildred, the customer does not write but clicks on pre-programmed answers and is led through a first problem diagnosis and troubleshooting process step by step. Video instructions are frequently posted as well making the service as the first point of contact quite useful. As technical problems can quickly become complex, however, the bot meets its limitations after a few questions and, upon requests, passes on to the classical means of customer service (https://www.messenger.com/t/freenetTV).

An absolutely extraordinary product for outbound marketing is shown by the advertising campaign of Kwitt, the payment system of the Sparkasse. In this prestigious example of creative marketing in symbiosis with AI and Facebook Messenger bots, it is shown how successfully this connection can be used. With Kwitt. Money is transferred from one mobile to another using the Sparkasse app, the only thing needed is the recipient's mobile number. With the "der Bote der Sparkasse" bot, a personal Moscow collection agency was created in a jiffy in a short, really funny chat. Return of debts guaranteed after that! (https://www.messenger.com/t/wirsindkwitt).

In contrast to most bot applications, the KLM bot is connected to the CRM system of the service centre and is thus able to escalate service cases the machine cannot process.

4.6.1 "Spooky Bots"—Personalised Dialogues with the Deceased

In 2016, there was a really unusual development of a chatbot: A memorial bot for a deceased friend. Eugene Kuyna, a bot developer of Russian descent from Silicon Valley, got the idea after receiving the devastating news about the casualty Roman Mazurenko. In defiance of all her ethical reservations, she collected thousands of lines of chats from other relatives and fed them into a neural network similar to how Amazon's Alexa or Apple's Siri were developed.

The results are both fascinating and scary. Many of Mazurenko's friends that spoke to the bot were staggered at the unique expression of Mazurenko's, which his bot had perfectly imitated in many places, Even his humour shines through at times. A friend once wrote to him, for example: "You are a genius!" and the bot replied quick-wittedly as Mazurenko would have: "And good-looking!"

Kuyna collected some log files of the chats in order to be able to get an idea of the outcome. She noticed that the bot listened more than it spoke. For many of the relatives, the benefit of the bot was therapeutic. They were thus able to tell it things they had always wanted to say. Many were able to bid their farewells to him in this way, a fact that would not have been possible without digital avatars. Yet, the effect can also turn into the opposite and the mourning phase of the relatives can be suppressed and extended.

The unusual and relevant example closely shows the possibilities that are open to all of us with this technology these days. Yet, the commercial weighing up of costs and benefits is at least just as important as the continuous further development of the technology. We are living in times where each one of us individually and society as a whole has to give some thought to a more responsible use of the new technologies to ensure a meaningful and profitable use. For, as many advantages as AI brings along, as with any kind of technology, they come with certain risks that are to be identified and avoided.

4.7 Maturity Levels and Examples of Bots and AI Systems

4.7.1 Maturity Model

The possibilities of implementation of bots are as diversified as the needs of the business and its customers. For a better overview, for degrees of maturity of chatbots can be differentiated (Fig. 4.7).

The first and lowest level is represented by chatbots without any access whatsoever to other data. Many bots that have been established in customer service until now can be categorised on this level. They secure basic communication, pick up the customer for the time being but soon reach their limits and pass the customer on to the next touchpoint.

On the second level, context information about the consumers is already used. For the duration of the interactions, the bot remembers the likes of

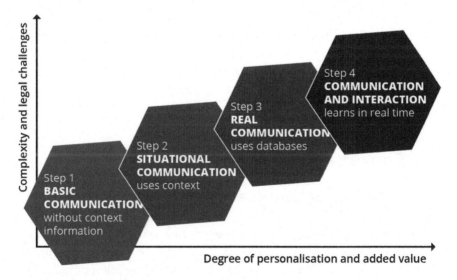

Fig. 4.7 Maturity levels of bot and AI systems

the customer's location or the products viewed in the shop and can make recommendations based on this. It is highly situational communication that offers a lot of potential for the customer journey, yet is not aimed at strong customer retention and an empathetic appearance of the system.

The next and third level is represented by a bot that has additional access to historical context information. It is the first level with real communication between the company and the customer. In the bot's memory, an internal database, besides previously purchased products there are also all of the customer's reviews and problems, which can be used accordingly.

An extensive personalisation is achieved with the fourth level. They are connected to the company's CRM system and add to it during customer interaction in real time. Digital butlers such as Alexa can be categorised here. They get to know their customers and act on behalf of the customer as a digital entity to place and order, for example. It is not only a communication system but there is actual interaction with the customer.

With the increasing degree of maturity, not only the complexity and added value of the bot increases, but also the legal challenges. Data protection implications of the application must be considered and weighed up, as the collection of customer data can be problematic. It depends on the scaling in this case, as well.

The use of surf context information, with the help of cookies for example, is usually unproblematic, even in Germany. In contrast to that, personal butler systems such as Amazon's Alexa, are being criticised by the public for collecting and analysing too much user information. Data protectionists are criticising the system on all leading media in this respect, which can turn the marketing of the product into a challenge. Likewise, the customer's inclination to use the system is also declining. In the worst case, trust in the brand can be shattered and a negative downward spiral of the customer review can be instituted. Enhancement effects and possible consequences must be weighed up carefully with the benefit.

4.8   Conversational AI Playbook

4.8.1   Roadmap for Conversational AI

Owing to the technological development and changes in the customer behaviour, e-commerce has developed over different levels of maturity in recent years. The challenge for companies is to recognise relevant technological and market trends and assess them accordingly.

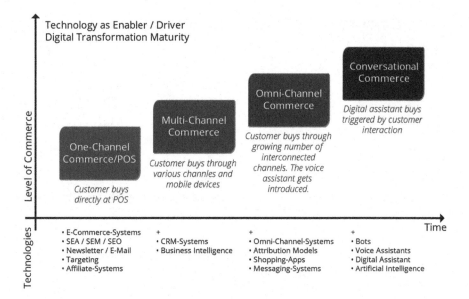

Fig. 4.8 Digital transformation in e-commerce: Maturity road to Conversational Commerce (Gentsch 2017 based on Mücke Sturm & Company, 2016)

Companies are currently facing the challenge of achieving the next level of maturity—so-called Conversational Commerce. This level of maturity seems desirable at present because current trends could revolutionise the sales sector. This means that those who proceed slowly with the implementation of Conversational Commerce could lose customers to competitors. On the other hand, companies could, for example, benefit from public attention by incorporating bots at an early stage (Fig. 4.8).

Thereby, the leap to Conversational Commerce does not represent a gradual, but a fundamental advancement of e-commerce. This is not only about another voice-controlled touch point. It is much rather about a new ecosystem which automatically initiates and coordinates ordering processes driven by customers and situations. Intelligent assistants either follow the instructions of consumers or recognise the need to take action by themselves, e.g. reordering of detergents or travel booking according to the appointments diary.

However, it is also decisive that the transition to Conversational Commerce is well thought out and planned. One possibility to do this systematically is the DM3 model presented in Part II AI Business: Framework and Maturity Model (Fig. 4.9).

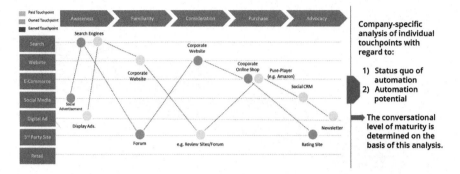

Fig. 4.9 Determination of the Conversational Commerce level of maturity based on an integrated touchpoint analysis (Gentsch)

4.8.1.1 The DM3 Model as a Systematic Procedure Model for Conversational Commerce

Each touchpoint has to be analysed both for itself and in conjunction with other touchpoints regarding costs, benefit and risk. This is the only way to derive the ideal current and future Conversational Commerce strategy. The general idea is to assess the trade-off between costs, benefit and risk. A high degree of automation of a touchpoint may have efficiency benefits but on the other hand also high costs and in some cases may lead to a suboptimal customer experience. A systematic comparison of costs, benefits and risks is thus indispensable.

Thereby it is not a matter of 0/1 decisions. Rather, a decision has to be made as to which degree of automation makes sense for which touchpoint (Figs. 4.10 and 4.11).

4.8.2 Platforms and Checklist

The question regarding the platform for Conversational Commerce is more of an operational question.

Companies should first decide on a platform their customers are already using. Facebook Messenger may be a good choice in many European countries and the USA because the number of users is very high there. If the circle of customers primarily consists of millennials (the generation born in the period around 1980–1999), Snapchat may be more suitable. WhatsApp, Viber or Line also dominate in many countries. If the target group is located predominantly in China, WeChat is the most suitable platform.

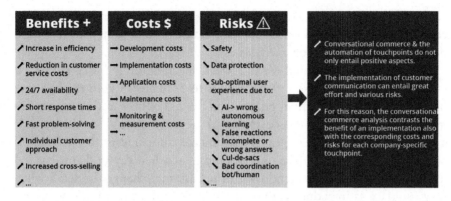

Fig. 4.10 Involvement of benefits, costs and risks of automation (Gentsch)

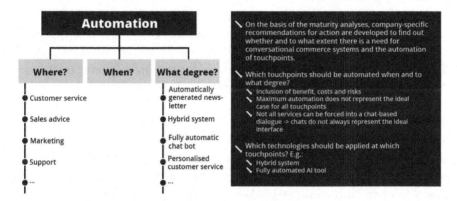

Fig. 4.11 Derivation of individual recommendations for action on the basis of the Conversational Commerce analysis (Gentsch)

The next step is to consider whether there are sufficient resources not only to create, but also to maintain a bot. This applies both in terms of professional expertise and personnel. Should the expertise not be available in the company, it is advisable to call on a partner for the technical implementation. But also the time and costs for maintaining the bot in the long term should not be underestimated. For although the bot is automated, time is needed to (a) promote the bot, (b) check the cases where the bot could not help, (c) measure customer satisfaction, and (d) constantly work on the improvement in the bot.

A further important aspect to be considered thoroughly is how the brand personality of the company can be maintained and promoted via Conversational Commerce. It is particularly important to communicate the values of the brand in the online chat, because these conversations have a very human touch. This implies that a consistent brand personality exists; in case of doubt, the brand personality should be created as quickly as possible before Conversational Commerce is used.

It is also essential that there is a clear, meaningful and well-studied use case for the use of chatbots. What goal is to be achieved with the bot and is it feasible—also in the initial phase? Will the use of bots lead to an improvement in the service for the customer? A negative example is the countless apps from which the user gains no advantage compared to the website. The customer uses every interface to the brand in another way so that it has to be investigated how the interaction with the customer changes in detail when a new interface is introduced. By analysing the current communication with customers, topics can be found that are suitable for using a bot. For companies it is generally worth it if the bots are implemented in stages and in clearly definable areas. In other words, the use of chatbots should be limited to those areas where it works especially well. The rest should be left to people until the technology is matured. This also increases customer acceptance. If the entire booking system of an airline, for example, is reorganised from the beginning, this may be very risky, because the probability that it will not run smoothly immediately is very high. Chris Messina emphasises that a bot should by no means be used for spam. In Conversational Commerce frustrated customers can strongly influence a company's success, because they interact with the brand in the same way as with a person. However, if it is possible to offer the customer a convenient, personalised and meaningful service, a company can considerably benefit from Conversational Commerce.

Checklist for companies

- What messaging platform are my customers using?
- Are there sufficient resources with regard to expertise and staff for long-term maintenance of the bots?
- Does my company have a brand personality and a strategy to communicate it in online conversations?
- Is the area where the bots are to be used clearly delimited and can the bots achieve the planned goals without disappointing customers?

There are numerous platforms for Facebook Messenger via which companies can set up a simple bot relatively quickly. Among them the following providers are recommendable:

- Chatfuel (http://www.chatfuel.com),
- wit.ai (http://www.wit.ai/, feat. by Facebook),
- and recast.ai (https://recast.ai/).

Those who want to create a successful bot need to pay special attention to the following points:

- Bots are something new. Their service should stand out against apps and websites and they should be unique, otherwise they may disappoint the users soon.
- Transparency is important. Nobody should simply replace human call centre agents by virtual assistants without designating them as machines.
- Personality: Users expect very personal service, even of bots. That is why many providers rely on bot names.
- Do not hide your bots but exercise bot marketing on all channels available.
- Interview your stakeholders to find out what they expect of bots.
- The more you know about your customers' needs, the better your bot can perform.
- The individualised approach through bot with customised content creates satisfaction and thus improves customer loyalty.

4.9   Conclusion and Outlook

4.9.1   E-commerce—The Deck Is Being Reshuffled: The Fight for the New E-commerce Eco System

Those in possession of the direct interface to the customer in the form of an own bot who know consumers' preferences and behaviour in all areas of life, determine information, advertising and purchases. Whilst consumers choose their favourites for themselves from a Google search hit list or an Amazon product search, bot recommendations usually reduce the recommendation to one product and one piece of information. The bot sovereignty thus replaces the active evaluation by the consumer.

That this struggle is highly relevant and profitable is demonstrated by the efforts of Amazon to gain control over the customer by means of the Dash Button and the DRS system under the pretence of convenience and by the many investments of Facebook and Microsoft in smart bot and messaging systems. The promising platform-independent messaging and bot system of the former-times inventor of Siri, Viv, was acquired by Samsung in October 2016 who is sure to interpret the platform independence differently now. Similar to the app economy that gained momentum through strong players such as Google and Amazon, an industry leader will also be required in the bot economy. A mere analogy with the app store will not suffice though.

A bot store would be bound in the application silos again and not do justice to the bot logic as a lubricant for holistic transactions.

The in-depth interlocking with the eco system of the customer offers companies unique possibilities of data acquisition and analysis. By centralising and monopolising the customer interface, companies can lull consumers in their commerce bubble on the basis of comprehensive preference and behaviour profiles.

Of course, companies have always analysed consumer data in order to align products and communication with the target groups and to be as profitable as possible. It is also completely legitimate for companies to act in line with their profit maximisation approach. With the only difference that companies and consumers no longer meet on traditional markets but the provider in sense internalises the market. Amazon ceased to be a dealer of products a long time ago and is now a smart eco system which intelligently collects, analyses and uses data to keep the consumer in its own commerce bubble.

4.9.2 Markets Are Becoming Conversations at Last

Markets are conversations reloaded: The postulation "markets are conversations" formulated in the Cluetrain Manifesto in 1999 is reinterpreted in the light of Conversational Commerce. Communication and interaction are increasingly controlled and determined by algorithms. The advantage is that conversations with companies demanded from the perspective of responsible consumers are now possible "at scale". Bots work in parallel at random in 24/7/365-mode. Obstacles to profitability and efficiency on the part of the companies were frequently an impediment to a personalised conversation. On the other hand, the pseudo-human dialogue means a loss of empathy and emotions. However, it is less a matter of the typical man-versus-machine battle but rather a matter of intelligently orchestrating and balancing both approaches.

Computerisation and algorithmisation in e-commerce is nothing new, of course. For a long time Google has been determining what products we see, Facebook's news algorithm defines our newsfeed and real-time bidding controls what advertising we get to see. What is new, however, is the extent of algorithmic coverage across the entire transactional value chain. In addition, the increasingly widespread mechanism of "added value for data" is reducing consumer sovereignty. As a result, the consumer sovereignty—determined significantly through the Internet—in the form of transparency and the possibility of rating companies and products visible for all to see is threatened.

A kind of bot sovereignty replaces consumer sovereignty. Due to the fact that present and future bots are offered in particular by the GAFA (Google/Amazon/Facebook/Apple) corporate world or are developed on their platforms by companies, consumers no longer have real sovereignty. GAFA bots offer convenience without having to pay for it directly. But the consumer then no longer makes a really sovereign decision.

It is to be expected that there will be a turnaround within Conversational Commerce in Germany in the course of 2017—following the examples from China and the USA. Presumably many online businesses will use bots to offer customers better and faster service. It is still unclear how far Conversational Commerce will expand across the different industries. It is clear that bots will be constantly improved and that the response and recommendation algorithms will be refined further. An optimal individual and automated interaction between customers and companies is to be expected in the long run, bringing about advantages for both customers and companies.

In the end an increasingly data-driven and analytical business will have to answer the question of the right balance between automation and personal interaction. It remains to be seen who will win the multi-billion dollar race in Conversational Commerce. The corresponding implications for consumers are equally fascinating. Will they be strengthened by the respective bot power in the form of digital assistants who know and adequately represent their actual preferences or will they rather become a puppet of the perfectly designed data and analytics eco system of the digital giants? Therewith, after the Internet, mobile and IoT, we are in the certainly most exciting phase of our digital transformation.

The USP and innovation of the presented DM3 model lie in the totality and stringency of the approach: The strategy does not remain at a high-level power point level but is systematically translated into appropriate measures and metrics. Instead of single individual activities an aligned and prioritised catalogue of measures is generated for successful Conversational Commerce—digital success with a system!

Note

1. Viv was taken over by Samsung in 2016.

References

Bager, J. (2016). Gesprächige Automaten. *C't – Magazin für Computertechnik, 24,* 112–114.

Brown, B. (2015). *Your New Digital Coworker.* https://blog.howdy.ai/your-new-digital-coworker-67456b7c322f#.jyo3j7r6q. Accessed 4 Jan 2017.

China Internet Watch. (2016). *WeChat Monthly Active Users Reached 806 Million in Q2 2016.* https://www.chinainternetwatch.com/18789/wechat-monthly-active-users-reached-806-millionin-q2-2016/. Accessed 4 Jan 2017.

Downey, S. A. (2016). *Bots-as-a-Service.* https://medium.com/@sarahadowney/bots-as-a-service-766287876ec6#.mhoa17re0. Accessed 5 Jan 2017.

Gangwani, T. (2017). *Hiring a Chief Artificial Intelligence Officer (CAIO).* http://www.cio.com/article/3157214/artifcial-intelligence/hiring-a-chief-artifcial-intelligenceoffcer-caio.html. Accessed 23 Jan 2017.

Gartner. (2013). Gartner Says by 2017 Your Smartphone Will Be Smarter Than You. *Gartner Press Release.* http://www.gartner.com/newsroom/id/2621915. Accessed 4 Jan 2017.

Gartner. (2015). *Gartner Reveals Top Predictions for IT Organizations and Users for 2016 and Beyond.* http://www.gartner.com/newsroom/id/3143718. Accessed 5 Jan 2017.

Gentsch, P. (2016). Die Bedeutung und die Rolle des CDOs bei der Digitalisierung von Unternehmen, DIVA-e Whitepaper Online.

Gentsch, P. (2017). Mit System Digital transformieren, DIVA-e Whitepaper Online.

Gentsch, P., & Ergün, C. (2017). Empirische Studie zu der Leistungsfähigkeiten von Bots, HTW Aalen in Kooperation mit diva-e, 02-2017.

Hammond, K. J. (2017). *Please Don't Hire a Chief Artificial Intelligence Officer.* https://hbr.org/2017/03/please-dont-hire-a-chief-artifcial-intelligence-offcer. Accessed 29 Mar 2017.

Hoare, G. (2014). Always Bet on Text. *Livejournal.* http://graydon.livejournal.com/196162.html. Accessed 4 Jan 2017.

iBusiness. (2017). Siri, Alexa, Cortana oder Assistant: Wer das Rennen der Sprachagenten gewinnt, Feb. 2017.

Libov, J. (2015). Futures of Text. http://whoo.ps/2015/02/23/futures-of-text. Accessed 4 Jan 2017.

Mckinsey. (2017). http://www.mckinsey.com/business-functions/digital-mckinsey/our-insights/intelligent-process-automation-the-engine-at-the-core-of-the-next-generation-operating-model.

Messina, C. (2016b). 2016 Will Be the Year of Conversational Commerce. *Medium.* https://medium.com/chris-messinga/2016-will-be-the-year-of-conversational-commerce-1586e85e3991#.e23seb2m9. Accessed 4 Jan 2017.

Nusca, A. (2017). *Yes, Your Company Needs a Chief AI Officer. Here's Why.* http://fortune.com/2017/01/05/artificial-intelligence-officer/. Accessed 5 Jan 2017.

Quoc, M. (2016). *11 Examples of Conversational Commerce and Chatbots in 2016.* https://chatbotsmagazine.com/11-examples-of-conversational-commerce-57bb8783d332#.fxn76d3ya. Accessed 4 Jan 2017.

Shevat, A. (2016). The Era of the Conversational Office. *Medium.* https://medium.com/slackdeveloper-blog/the-era-of-the-conversational-offce-e4188d517c64#.jwbb8293p. Accessed 4 Jan 2017.

Strauß, R. E. (2016). Künstliche Intelligenz Goes Marketing. Absatzwirtschaft. Sonderausgabe zur dmexco, 34.

Van Doorn, M., & Duivestein, S. (2016). *The Bot Effect: 'Friending Your Brand'.* Report. Applied Innovation Exchange, SogetiLabs.

Vouillon, C. (2015). Slackbots. *Medium.* https://medium.com/point-nine-news/slackbots-9144feee6f6#.hi5qc32jn. Accessed 4 Jan 2017.

Part IV

AI Best and Next Practices

5

AI Best and Next Practices

5.1 Sales and Marketing Reloaded—Deep Learning Facilitates New Ways of Winning Customers and Markets

Andreas Kulpa, DATAlovers AG

5.1.1 Sales and Marketing 2017

"Data is the new oil" is a saying that is readily quoted today. Although this sentence still describes the current development well, it ides not get down to the real core of the matter; more suitable would be "artificial intelligence empowers a new economy". The autonomous automation of ever larger fields of tasks in the business world will trigger fundamental economic and social changes. Based on a future world in which unlimited information is available on unlimited computers, ultimate decisions will be generated in real time and processes will be controlled objectively. These decisions are not liable to any subjectivity, information or delays.

In many sectors of the economy, e.g. the public health sector or the autonomous control of vehicles, techniques of artificial intelligence (AI) are applied and increase the quality, availability and integrity of the services offered. The same development can be observed in the field of sales and marketing. Today, companies no longer allow themselves to be recorded by turnover, commercial sector and other company master data. Presence and active

© The Author(s) 2019
P. Gentsch, *AI in Marketing, Sales and Service*,
https://doi.org/10.1007/978-3-319-89957-2_5

communication on the Internet, be it the website or in social networks, today belong to a company's everyday routine. The efficiency of a sales or PR campaign heavily depends on the choice of companies and people to be addressed. Are they interested in the subject? Is this a well-chosen point in time? Has the company just concluded a contract with an innovative CMS provider, or is an outdated stack still being used? Classical sales and marketing approaches define target groups by way of simple selections or segmentations. Companies are selected on the basis of commercial sectors and sales margins and transferred into the sales process.

Prior to the first call by the sales team, little can be said about the probability of the conversion with this approach. There is neither data nor a method available to make a forecast about whether the prospective customers can really be won over as a customer in the sales funnel. Yet, for an efficient and agile sales process, having extensive and up-to-date data is crucial. The establishment and development of individual leads in issues of the topics they focus on, their sales forecasts and their digitality are crucial for successful communication. Accordingly, an ideal system should make a sure prediction as to which prospective customer will be the next to sign a contract. This way, the sales team can achieve the maximum conversion rate.

The high complexity of the data and the high dynamics this data underlies are a typical field o application for deep and machine learning algorithms. In the following, I will illustrate how these are applied to the field of automated lead prediction.

5.1.2 Analogy of the Dating Platform

> Tell us your customers with the highest sales and we will predict who your next successful customers will be. (Kulpa 2016)

In principle, lead prediction can be easily compared with a dating platform. In comparison with a simple assumption about which products go well with a company, lead prediction learns new information from every new customer to, in turn, predict better customers. The predictions become more reliable and precise from the interaction and the feedback resulting from it (Fig. 5.1).

In comparison with a sales rep, who avails of a subjective and limited view of the companies in the sales pipeline and the market itself, lead prediction approaches use a wide spectrum of data from various sources, which is merged to an ideal outcome in a highly topical and highly dimensional deci-

ONLINE DATING PLATFORM

Online Dating in the post-processing phase

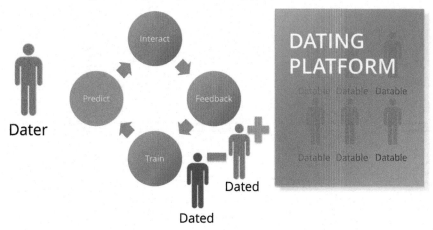

Fig. 5.1   Analogy to dating platforms

sion-making process. The features used can be divided into different groups and they consider various aspects and properties of the suspects.

5.1.3 Profiling Companies

Under many different aspects, a comprehensive picture of every potential lead is generated. The task at hand is the complete recording of the current state and an estimation of the development of the company. This covers the classical master data, an exact classification of the activity and an estimation of the development of the company in its sector (Fig. 5.2).

5.1.4 Firmographics

Firmographics contain traditional company data that is taken from the company registry (name, location, commercial sector) and extended by further indicators such as turnover and number of employees. The commercial sectors are a classification of the activities of companies that were published by the Federal Statistical Office in 2008.

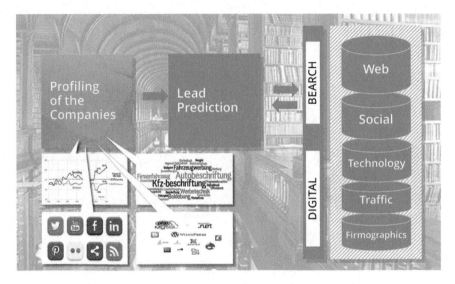

Fig. 5.2 Automatic profiling of companies on the basis of big data

5.1.5 Topical Relevance

Thanks to the dynamic identification of the subjects from the website, in comparison with the commercial sectors, lead prediction achieves a very accurate thematic classification and localisation of the company. I addition, these tags have high topicality and new trends quickly become visible. In comparison with the commercial sectors, instead of commercial sector software development, a company is given the tags app development, big data or machine learning.

Word2Vec is used for the thematic classification of companies. Word2Vec was released by Google in 2013 and is a neuronal network, which learns the distributed representations of words during training. These vectors have astounding properties and abstract the semantic meaning in comparison with simple bag-of-word approaches. Words with similar meanings appear in clusters and these clusters are designed such that some word relations such as analogies can be reproduced under the application of vector mathematics, as in the famous example: "king − man + woman = queen".

Via the Word2Vec presentation of texts, operations can be mapped; the relationship of Apple to smartphones is identical to the relationship of dell to laptops.

5.1.6 Digitality of Companies

The digitality of a company shows how far the company has completed the process of digitalisation. Various aspects of digitality are included in this score: the technology of the website, the visibility of the company on the web, the ad spending and SEO optimisation and the degree of innovation of the business model. On the basis of this score, companies can be easily segmented depending on the degree of their digitalisation. Both young start-ups and established companies in the e-commerce sector are distinguished by a higher-than-average digital index, whereby more traditional business sectors reveal a rather less distinct degree of digitality. Table 5.1 shows the individual dimensions of the digital index (Fig. 5.3).

5.1.7 Economic Key Indicators

Key indicators from the investor relations environment are determined for every company.

- Development of the staff: A stable or a growing number is a sign of a positive development of the company.
- Consumer activity: What is the situation in the individual commercial sectors and how is the development estimated?
- Does the company pursue technological trends?

Table 5.1 Dimensions of the digital index

Dimension	Attributes
Tech	Hosting, CMS, Server, Frameworks, Widgets, JavaScript, CND, Analytics, etc.
Traffic/reach	How much traffic does the site generate? How many users see the site? Unique visitors, page views
Mobile	Mobile readiness: Are the offers also designed or optimised for mobile devices?
Search	SEO & advertising: Ads spending and SEO optimisation available?
Social	Social media comprises: • Social media readiness: How many channels is the company represented on? • Social media activity: how active is the company on the social media channels?
Connectivity	How well is the company networked?
Quality	How does the user perceive the quality of the website? How fast does the site load? How well-written re the texts?
Innovation	How innovative is the company's business model?

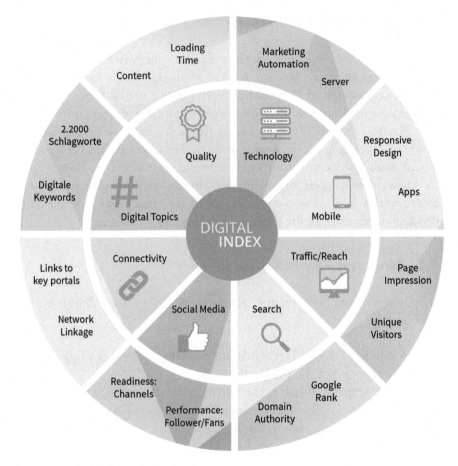

Fig. 5.3 Digital index—dimensions

Base on this spectrum of data, which is available in high topicality, the lead prediction generates a presentation that summarises all aspects of the company in a 360° perspective.

5.1.8 Lead Prediction

The characterisation of the entire companies that should be used for lead prediction is an essential step. On the basis of this generic customer DNA, further companies are identified that have the same DNA (Fig. 5.4).

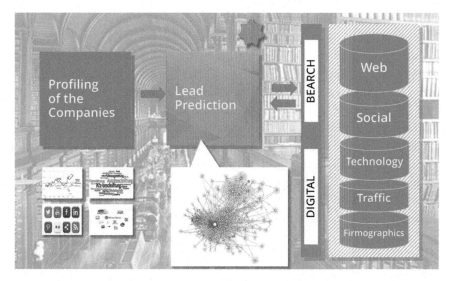

Fig. 5.4 Phases and sources of AI-supported lead prediction

5.1.9 Prediction Per Deep Learning

Deep learning is a subject that is causing quite a stir at the moment. In principle, it is a branch of machine learning that uses algorithms to recognise objects, for example, and understand human speech. The technology is in principle a revival of algorithms, that were popular from the beginnings of AI: Neuronal networks. Neuronal networks are a simulation of the processes in the brain whereby neurons and the specific fire patterns are imitated. The real innovation is the layering of various neuronal networks which, in combination with the essentially greater performance of current computers, led to a quantum leap in diverse sectors of machine learning.

The classifier for the prediction learns a generic DNA on the basis of profiling the successful customer relations, which is projected onto the entire company's assets. The prediction of the optima leads can be understood as a ranking problem. The lead with the highest probability of a conversion should be in first place in the sales pipeline. In principle, it can be understood as a classic regression task where the probability of conversion is to be predicted. Thus highly suitable is a gradient boosted regression tree, also called random forest.

5.1.10 Random Forest Classifier

The algorithm gradient boosted regression trees, also called random forests, belong to the ensemble learning methods This classifier uses an ensemble of weak regression trees that have a low hit quota when considered in isolation. The quality of the prediction can be improved significantly when various trees are trained with different parameters or samples. The results of the individual trees are aggregated to a total result which then enables a more balanced and high-quality prediction. The so-called bagging triggered a boom of the traditional regression trees. As aggregation, either a majority vote or a probability function is chosen (Fig. 5.5).

The lead prediction generates high-conversion leads because

- The entire spectrum of information available about a company is integrated into the decision-making;
- The data is highly topical and without bias;
- The random forest is capable of abstracting complex correlations in the data; and
- The method learns iteratively from the interaction with the sales team.

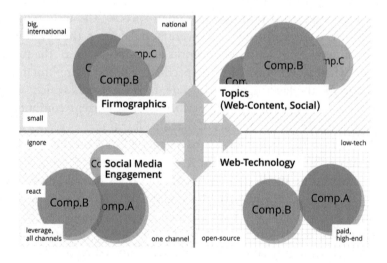

Fig. 5.5 Lead prediction: Automatic generation of lookalike companies

The choice of leads is the first step in the sales process; the second one is to find the ideal point in time for addressing them.

5.1.11 Timing the Addressing

The right addressing, the right occasion and the right point in time—good content marketing demands each individual aspect to be as successful as possible. Various studies have shown that important purchase decisions are made at certain occasions in life. When marketing and acquiring new customers, a well-chosen point in time of addressing them is essential for success or prospect of conversion. In a sales team, this is typically done intuitively on the basis of experience. How is this decision made if this knowledge has not yet been gathered? We have developed our own approach.

5.1.12 Alerting

We scan the Internet for signals and this way, we are informed about economic changes in companies. Any mention of companies is analysed and the impact they have is evaluated and whether they reveal a positive or a negative development. A rapidly increasing number of complaints to a non-responsive customer service can be an indication of internal problems within the company. News, blogs, social media and the website are a highly topical source of information about the condition and development of a company. Scheduled relocations, structural changes, expansion strategies or profit announcements, for example, are quickly visible and are a sign of a positive or negative development of a company. On the basis of these "early signals", statements can be made about how probable a company will react to being addressed at the current point in time.

Alerting openly scans the Internet and crawls cyclically websites and social media channels for content snippets containing information about a company. These snippets are the potential alerts that are filtered and aggregated according to significance down the line. In the first step, the probability of a company being mentioned in the given text is determined. Sequence learners, which make a decision based on the lexical similarity and the context of the word as to whether the mention refers to a company or not, are used for this purpose.

In the second step, a deep learner decides whether the validated snippets on a company trigger an alert or whether they are a part of daily background noise. To this end, a model is trained on the basis of historic text data and corresponding share developments, to recognise correlations between snippets and the development on the stock market. The time lag between alert and real change "lag" is automatically learned by the system. Recurrent neural networks, in comparison with other approaches on the basis of a "sliding time window" in combination with a classical regression, do not have the limitation of the finite number of input values.

Subsequently, the system is in a position to make its own predictions about the profit development of a company. These indicators are used in lead prediction to choose those companies among those with a very similar DNA that, at the current point in time, are most probably interested in an evaluation of the business activities.

5.1.13 Real-World Use Cases

5.1.13.1 Company: Network Monitoring

The spectrum of customers of these companies is diversified. Many of these companies are located in the environment of information technology and offer server hosting, for example. O the other side of the spectrum somewhat exotic companies emerged, such as operators of large production plants, silos, chemical production plants, etc. A manual evaluation of the leads from the lead prediction turned out to be difficult, meaning that we decided in favour of A/B testing. In the sales process, the leads that were predicted by lead prediction scored higher than average and generated a 30% higher conversion rate.

5.1.13.2 Company: Online Shop for Vehicles Construction and Industry

Two predictions were made with this project. The first one was aimed at the regular customers, the second one at customers that did not belong to the sales team's general target group, but were acquired by chance instead. The aim was to increase the market of this so-called alien group, in order to enter a market segment that had not yet been defined in detail. The conversion rate improved by 40% in the classical segment; in the new segment, an increase by 70% could even be measured

5.1.13.3 Company: Personnel Service Provider

A clear lift can be recognised with this prediction case. Via classical list generation, seven appointments used to be generated from 700 telephone calls; that is a conversion rate of one percentage. On the basis of the leads determined per lead prediction, there were nine appointments from 300 telephone calls; that is a conversion rate of three percentage. That is a significant increase. However, it should be kept clearly in mind that this is quite a small sample.

5.2 Digital Labor and What Needs to Be Considered from a Costumer Perspective

Alex Dogariu, Nicolas Maltry, Mercedes-Benz Consulting

In this use case by Mercedes-Benz Consulting the experience from thousands of real-life customer/Digital Labor interactions as well as numerous customer research studies on Digital Labor are summarised and the necessity for a centralised platform approach towards Digital Labor is elaborated on.

The landscape for customer management, customer experience, CRM and customer service is changing rapidly due to the evolution in AI and its growing adoption in real-life use cases. One particularly rapid growing application of AI are Chatbots and Digital Assistants in customer interaction. The trend towards the automation of work and the associated savings potential in terms of workforce creates also great concerns, skepticism and fear among the workforce and thus also in the public perception. In addition to contributions on the opportunities of digitalisation, AI and automation, there is always talk about demands for regulations and guidelines. For example, trade unions speak of "job killing by artificial intelligence". The integration of human and AI-based labor is thus very important, but won't be further discussed in this use case.

"Mercedes-Benz Consulting analysis of various contact centre data in the automative sector revealed that 80% of customer inquiries are repetitive and are rather simple in nature". These 80% can thus be automated through digital labor. We usually refer to this as the "Fat Head and Long Tail" approach (Fig. 5.6).

When looking at current customer service operations, we can clearly identify plenty of opportunities for digital labor. The ordinary customer service opening hours range from 8 a.m. to 8 p.m. This frustrates customers, espe-

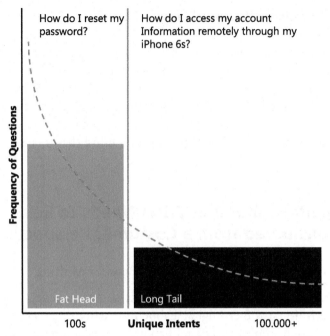

Fig. 5.6 Fat head long tail (*Source* Author adapted from Mathur 2017)

cially if they have a specific concern or a problem with a purchased product or service and want to get it dealt with immediately. Analysis of for example live chat data has shown that most live chat requests happen between 8 p.m. and 11 p.m. on a daily basis and mostly on weekends on a weekly basis. The use of Digital Labor is very useful here as it is available 24 hours a day, 7 days a week.

As machine learning, natural language processing (NLP) and robotic process automation evolve, Digital Labor will also be able to take on increasingly complex tasks. For example, answering more difficult customer inquiries in a personalised way or perform business tasks. The idea behind this is that digital labor will recognise the needs of the customer in advance and based on previous behaviour, decisions and existing preferences, proactively engage in conversations to help customers and promote products and services.

Currently, the clear majority of customer interactions with digital labor include the ability to escalate to real agents or customer service staff. The already mentioned agents are a valuable, expensive and limited resource whilst customer requests to contact centres are steadily increasing. Therefore, automatisation and self-service turns out as a promising option for contact centre managers. Software that can reply to requests in a natural way (e.g. a Chatbot) is a strategically important chance to handle rising costs and customer expectations (Aspect 2017).

An analysis of Frost and Sullivan in 2014 (as cited in Accenture 2016) found that the total costs of ownership for contact centres increase by a compound annual growth rate of 14% in a five-year period regardless of size. Furthermore, they found that the total cost of ownership double in a period of 5 years due to an increase in digital services and customer requests. The biggest cost driver of contact centres are the fixed human labor costs with 75%. One of the main pain points of contact centres is the agent fluctuation in the first six months. More than one in three agents (37%) quits the job or gets fired. Accenture calculates in a high-level business case with a savings potential of 15–20%. They use "cost per interaction" in an ordinary contact centre at the level of 2.52€ and at a digital assistant/chat bot with 0.30€ (Accenture 2016). Digital Labor provides a huge opportunity to reduce the massive salary costs. The potential cost savings of customer service represent-atives through Digital Labor in the US are $23 billion, with total salary pay-ments of $79 billion (McKinsey 2014, qtd. as cited in Beaver 2016).

The vision of full automation in customer service consists of several implementation steps (Fig. 5.7), which should be approached gradually. Here, it is important to understand that these steps must first be checked for their own functionality and practicability. Also, considering the differ-ent data bases and existing CRM systems in companies, training of cognitive systems and the limitation of financial resources required for full automa-tion, implementation can only be step-by-step. The vision of a 100% auto-mation is not feasible.

Developers of Digital Labor obviously want to implement personalities in their assistants in order to engage in an emotional manner with their coun-terparts. This serves to encourage users to feel a certain sympathy towards Digital Labor employees and thus achieve a certain degree of friendship (customer loyalty). But if you speak with a Digital Labor entity you behave differently. If you know that you are communicating with a system, there are no more human and ethical inhibitions. Moreover, there is no fear of being judged and one speaks much freer than with a real person. That explains

Fig. 5.7 Solution for a modular process (*Source* Author adapted from Accenture (2016))

why Digital Labor employees relatively often get insults, because one has no inhibitions to violate a software program (Arbibe 2017).

There has been little research on the topic of "trust in Chatbots" so far. Current and available publications with first survey values were used as a basis nonetheless. In the field of trust, data protection and the perception of privacy aspects by users also play an important role when communicating with a Digital Labor employee. Personal data (name, surname, gender, age, nationality, etc.) as well as contact data (address, telephone number or e-mail address) or administrative data (insurance or bank data) are highly relevant. In addition, customer data (contract length, current car, lease agreement data, etc.) are always queried in business context. There is always the risk of data leaks. Famous examples such as Yahoo or Deutsche Telekom are mentioned at this point.

"Digital trust has been shaken by a proliferation of malicious content and data breaches, which has significant consequences for brands that use these kind of Digital Labor platforms" (Elder and Gallagher 2017). In addition, due to new data protection regulations of the European Union (start 2018) the user will act even more sensitive here. This, of course, also depends on cultural circumstances (Arbibe 2017). For example, German customers are more sensitive than American ones, which is why the aspect of data protection related to Chatbots and customer trust is of very high relevance.

Mindshare and Goldsmiths Institute (University of London) defined several important aspects for building successful Chatbots. Besides elements like tone of voice/brand alignment, definition of the Chatbots role and the guideline to make people feel human it is important that companies/brands focus in establishing trust to rise Chatbot acceptance among customers.

"One of the key challenges any brand will face in building a bot is the issue of trust. It lies at the core of good customer service (our research found that 76% of people say trust is key to good customer service) and it has to be established before users can be expected to take part in deeper engagements" (Mindshare, n.d., p. 15). Furthermore, a very interesting aspect is that people prefer to provide sensitive information to a Digital Labor employee rather than to a human person. Mindshare and the University of London state: "Only 37% say they are happier to give sensitive info over the phone to a human than to a Chatbot. For 'embarrassing medical complaints', twice as many people prefer talking to a Chatbot than a human than for 'standard medical complaints'" (Mindshare, n.d., p. 15). Another important fact in the context of trust and Digital Labor is transparency.

"75% agree that 'I'd prefer to know whether I'm chatting online with a Chatbot or a human'" (Mindshare, n.d., p. 16). User want to know with whom they are having the conversation, because trust can be undermined by uncertainty. Also, the escalation to a real agent plays an important role for building trust. "79% agree that 'I'd need to know a human could step in if I asked to speak to someone'" (Mindshare, n.d., p. 16). Especially if a Chatbot is built to generate leads trust is essential.

To ensure the success of Digital Labor in customer service it is essential to regularly monitor the relevant Key Performance Indicators (KPIs). One example is the "Task Completion Rate" which means the percentage of successful completed tasks by an artificial intelligent unit.

5.2.1 Acceptance of Digital Labor

The key results of studies performed by Mercedes-Benz Consulting within the automotive sector provide initial insights for developers, project managers, researchers, and companies involved in Digital Labor projects.

5.2.2 Trust Is the Key

The knowledge gained in this work on the very influential factor trust has consequences for the further design of the tonality and contents of Digital Labor employees. It is important that Digital Labor employees trustfully communicate with customers. Thus, a trustworthy choice of words is very important at the beginning of the conversation, but also at critical points in a service session. Transparency and clear communication that concerns the storage and use of a user's data during a session is equally important. At the beginning, the Digital Labor employee could ask the customer if it should provide further information on data protection. Alternatively, it should be checked, if not all existing information on data protection as modelled questions can be represented by the Bot. If a Bot could answer these questions, the trust relationship might be positively impacted. It is a well-researched phenomenon that people reach a higher confidence level when communicating with a Bot that has a face or persona. This principle should be transferred to all Digital Labor employees. Here, we must take comprehensive measures because this our research shows that customers will not use Bots if they do not trust them with their personal data—despite a conversational interface that imitates human communication. Especially when Bots make orders, payments or financial transactions, trust plays a major role and has a huge impact on user adoption levels.

If companies offer personalised Digital Labor assistants to their customers in the future, who have as much information as possible to anticipate the preferences and wishes of the customer, a high level of trust is the basis. But also, if a digital assistant is available across many devices (smartphone, smartwatch, in-car assistant, Live Chat), this is of high relevance. In essence, Digital Labor employees should be designed in such a way that the customer feels the entire communication as a partnership between man and machine. Think or Wall-E instead of Skynet.

5.2.3 Customer Service Based on Digital Labor Must Be Fun

In our customer studies it could be confirmed that hedonic motivation is an important factor for customer engagement with Bots. Thus, when designing customer service Bots, one should take into consideration that gamification is important. This can be applied to the way a Digital labor employee answers (e.g. personality, jokes, chit chat), its virtual appearance, and built in games like text adventures or quizzes and so forth. Media agencies have a whole array of options to spice up your Digital Labor employee. Cognitive services offer many options, like picture recognition games for example. The Google Assistant has some interesting games built in, too.

5.2.4 Personal Conversations on Every Channel or Device

No surprise here. Customers want an omni-channel experience that is personalised. This means that the Digital Labor employee should recognise customers, greet them personally, remember the last conversation and know the customer journey. What surprised us was that customers do not really care about social media influencer's opinions or ad campaigns. They trust their own experience with a Bot. At Mercedes-Benz Consulting we made sure that our Bots know the vehicle model of a customer and even differentiate between left and right hand drive. Furthermore, we incorporated system and context variables and episodic memory in our Bots. User input is even used to personalise the content of a Bot. The goal is to develop data-driven user journeys like Netflix does it.

5.2.5   Utility Is a Key Success Factor

It could be confirmed that customers would like to interact with Bots, if these can make things easier for them. However, it should be considered that the performance of ordinary Chatbots is very limited and the expectation of a Chatbot's capabilities are often not met. In order to really help customers get their issues resolved, full process automation is key. Therefore, backend integration with CRM systems, transactional systems, customer data bases, etc. must be done. One of the key projects is to build data lakes and make all systems available via one API. Bots can then easily integrate with backend processes. This provides customers with a true one-stop solution. Current use cases range from changing personal data, booking a test-drive, making a service appointment to altering lease contracts via Bots. We constantly add new functionalities to our Digital Labor in order to increase the utility for our customers.

5.2.6   Messaging Is Not the Reason to Interact with Digital Labor

With Facebook messenger, WeChat, WhatsApp and other messaging services growing rapidly in the number of users across the world, making Bots for these channels seems obvious. The reason for text interaction, is the high convenience factor in asynchronous communication. One can communicate at anytime from anywhere and answer whenever he or she wants to or has time to do so. This high convenience factor does not automatically mean that customers accept Bots or Digital Labor just because they are available on these channels. In our customer studies this conviction could not be confirmed. Customers interact with Digital Labor because they expect to get their job done. The fact that the interaction is simple and intuitive and can be done via messaging services is given and not considered something special.

5.2.7   Digital Labor Platform Blueprint

When first starting out back in early 2016, the major use cases realised by Mercedes-Benz Consulting were FAQ style Chatbots, with little backend integration and a limited scope (e.g. frequently asked questions about e-mobility). User interaction volume and the level of satisfaction with the bots were rather low.

Looking back ten to fifteen years, even then FAQ style Bots existed, sometimes with weirdly looking avatars. However, this trend faded away quickly since users tended to prefer talking to a human being to get their inquiries handled or as one of the senior executives at Mercedes-Benz Consulting put it "get the job done". The main issues back then and with FAQ style Bots were: (1) Lack of context sensitivity (e.g. remembering previous user input, channel, current information domain), (2) No automation of business processes (e.g. change of personal data or ordering of a brochure), (3) Limited personalisation (e.g. recognition of user), (4) Limited scope of each bot (e.g. one bot for each information domain), (5) No omni-channel customer experience (e.g. transition from one channel to another leads to loss of context), and (6) Failure to set up hybrid bots (ability to hand-over to human agent at any point in time). Lessons learned also from the app landscape at many automotive companies, where hundreds of single purpose apps have horrible ratings and very low download rates.

From a company perspective, Chatbots and Digital Assistants were quite expensive and time-consuming to develop, as each use case required a whole content team to script the deterministic dialogues and answers and developers needed to integrate each Bot with new channels, databases as well as train the NLP unit and so forth from scratch.

When most companies still look at Chatbots as a onetime effort, Chatbots and Digital Assistants could represent the future of our workforce. Content needs to be updated constantly, information domains extended and business processes automated in order to stay relevant and interesting for our customers. That is why we extended the concept and coined the term digital labor within our domain of work.

With this in mind, Mercedes-Benz Consulting made sure from the beginning to strive towards a centralised digital labor platform for each of its clients. A digital labor platform is defined as shared service platform that serves as a basis for all client facing as well as internal Chatbots across all divisions. This way backend interfaces, frontend channel integrations and supporting tools like monitoring dashboards can be reused. Furthermore, Chatbot content can easily be reused or adapted: among others this might be the ground-truth, dialogue nodes, answers, and rich media content and so forth. Finally, most cognitive services pricing models are based on consumption with decreasing fees depending on the total number of API calls. Hence, a platform will always have lower costs per API call due to the combined volume of all Chatbots.

The advantages of such an approach also affect the customer experience tremendously. Customers can now switch between channels and the Bot aka digital labor employee still remembers the last conversation, user input, personal data and context, making any interaction seem more personalised and natural. By bundling all Chatbots under one umbrella—the "Agent Hub" or "Meta-Bot", customers can get a multitude of issues resolved by one entity. An intelligent algorithm routes incoming requests between the skills without the customer even recognising. This makes a skill activation as some might know it from Alexa or Google Assistant needless. Automatically any new dialogue, answer or business process one of the bot learns makes the whole system feel smarter. Thus the knowledge and skills are extending continuously, delivering real value to the customer by fixing more and more of his or her problems and handling inquiries automatically 24/7.

As depicted in Fig. 5.8, Mercedes-Benz Consulting Digital Labor Platform Blueprint consists of four distinct layers: (1) Connector Hub, (2) Content Hub, (3) Services Hub, and (4) Data Hub.

Since digital labor needs to be managed just like any other employee, we introduced a task manager and kind of a ticketing system to deal with workflows. A vendor-independent service orchestrator, often referred to as middleware, has been set up to connect all cognitive services like natural language processing or image recognition with each skill depending on the type of data being processed. Additionally supporting tools have been added one by one to the platform, as backend workflows needed to be automated or interactions stored in a knowledge management system for future developments of skills.

One example in the automotive sector, where Digital Labor is being employed is the virtual service desk (Fig. 5.9).

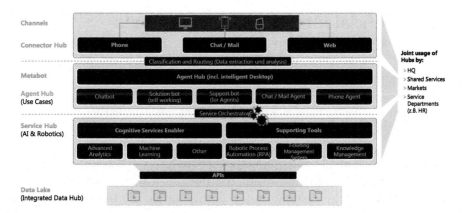

Fig. 5.8 Digital Labor Platform Blueprint

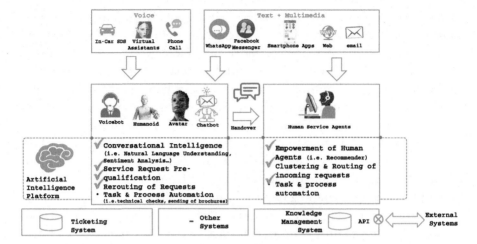

Fig. 5.9   Virtual service desk

Advantages in this Digital Labor scenario range from 24/7 self-service availability to immediate response times for customers. From a business perspective, the following advantages could be realised: significant cost savings by virtualisation of 1st level support, improved handling of peak times, reduction in call routing rate by prequalification and clustering, efficiency gains and cost savings by task/process automation, empowerment of human agents (i.e. recommendation for next best activity, answer, offer).

Based on the positive customer feedback, we are sure to extend our Digital Labor efforts into every functional unit and increase the depth of process automation in the coming months and years.

5.3   Artificial Intelligence and Big Data in Customer Service

Prof. Dr. Nils Hafner

5.3.1 Modified Parameters in Customer Service

Since the launch of the smartphone, digitisation has drastically altered customer service across many industries. In principle, it is now possible to know much more about the customer prior, during, or shortly following the contact service itself than even just a few years ago, and to accordingly treat them based on their individual needs. This offers an interesting prospect for improved profitability in customer service. From this point of view, the present article shows which possibilities of the intelligent use of various types of information from different sources and in a variety of formats (big data), as well as through the application of AI and machine learning, result in client contact.

Based on the explanation of big data and AI (2. A buffer's guiding to AI, Algorithmics and Big Data), it is clear that the application of the concepts can extend to a wide range of different service problems in various sectors. In order to make a classification that is useful in customer service, a proven instrument of strategic control for service incidents is relied upon. This would be the Value Irritant Matrix presented by Price and Jaffe (2008) displayed in Fig. 5.10.

On the one hand, the company subsequently considers whether it is interested in establishing contact with the customer from a service point of view, since it would provide them with knowledge about their products and services, thereby generating ideas for savings as well as the opportunity arising through the contact, of either selling other products or services or not. On the other hand, the customer's perspective on the contact services is systematically taken into account. This concerns whether the customer is truly interested in a personal contact to have his questions answered,

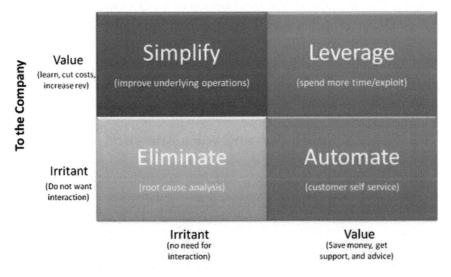

Fig. 5.10   Value Irritant Matrix (*Source* Price and Jaffe 2008)

receive advice, and, ideally, save money, or whether he does not see any need to make a contact with the company and would find any such contact bothersome.

The fundamental idea is that a company should assess where both customers and companies demonstrate an interest in making personal contact. It is only in such cases that valuable discussions take place. If there is a divergence of interests, where the customer has a high interest in solving a problem whilst the company regards the contact as a mere additional cost, the contact should be automated. This is of particular interest in the case of repetitious client questions. In this context, it is often a question of understanding how products and services function, also known as Self-Service. The same applies to the reverse case, in which the company is dependent on the customer to disclose certain information through the established contact, when it concerns a check-in or an e-mail confirmation, for instance. Such contacts are often regarded as bothersome by customers. In this case, the necessary customer notification contacts, such as for a check-in or partial contacts, are preferably simplified.

In the past few years especially, digitisation has led to many possibilities and ideas for, on the one hand, automating or simplifying more contacts, and, on the other hand, improving the customer experience in the so-called leverage quadrant. This always takes place under the premise of maximising benefits for companies and customers alike.

In order to demonstrate the contribution of big data and AI to this benefit maximisation, the following three areas of application shall be described:

1. Voice Analytics
2. Chatbots and Conversational UI
3. Predictive Servicing

5.3.2 Voice Identification and Voice Analytics

As a data source, the use of human language in targeted customer treatment has been on the rise over the past few years. Two possible applications exist in customer service; on the one hand, customer language identification. The potential is great particularly in industries where customer identification is required prior to the interaction for reasons of safety or proof of identity, since few customers remember the defined security passwords or wish to provide customer ID numbers or birth dates. As regards the Value Irritant Matrix, this has to do with streamlining contact with customers.

In this respect, various companies are already using the so-called voice-print. This voiceprint is a file containing the characteristics of a particular voice, such as frequency, loudness, speed, etc. However, no conversation content nor parts thereof are recorded. With a voiceprint, a person's identity can be authenticated with an accuracy rate of over 99 percentage. Additionally, identification is established on the basis of data that cannot be acquired with fraudulent intent. This can also be a means against social engineering attacks. In such attacks, fraudsters pretend to be customers in an attempt to obtain sensitive data.

Such services have been offered by providers such as Nuance and Nice for some time. However, only 7% of all contact centres use speech-based information to identify or even analyse call content. This is revealed by the 2017 Service Excellence Cockpit survey results (see the development of the Egle et al. 2014), in which over 180 European contact centres participated. Therefore, for many companies, this constitutes yet another potential for differentiation, as biometrical identification reduces the duration of the call for customers and companies, thus allowing customers to obtain their desired competent response more quickly (Service Excellence Cockpit 2017).

The great potential here lies in linking speech analysis and machine learning. This is demonstrated by the company Precire Technologies from Aachen, Germany. The founders of this company claim to have deciphered human speech, and this through the results of psychological studies and the use of big data technology. Recorded customer interviews can provide general statements on the communicative impact of a language, on emotions, on the personality and linguistic capacities of a person, but also on the motives and attitudes of groups or individuals.

In the contact centre field, this is particularly relevant for the interaction between employees and customers. Once the tool described has been taught in the sense of machine learning, and has thus understood what constitutes a "successful dialogue" from the company's point of view, the true customer (and employee) satisfaction can be measured and analysed at prior, during, and following a call. As such, the company saves the extra step of post-call surveys and can compile individual training programs on the basis of objective measurements. This is how both employees and managers receive coaching benefits from the increasing digitisation of customer service. The ultimate goal of the analysis is that calls become both shorter and more successful in terms of customer satisfaction as well as with regard to cross-selling and upselling.

All of these effects add up to interesting business cases, as shown by an investigation of two contact centres. The use of a speech analysis software paid off within just 5–7 months (Hafner 2016).

The automated customer satisfaction surveys can be seen as a real enhancement to today's NPS "benchmark", which, according to Service Excellence Cockpit is already applied in 40% of all contact centres. The customer then evaluates the quality of the relation on the basis of a scale of 0–10 answer to the question "Would you recommend us?" This evaluation is subjective, may be subject to political considerations, and is based on longer-term experiences (Reichheld 2006). The same bias is present in the rating of the question "Would you recommend us on the basis of the previous interaction?" and cannot be regarded as an expression of satisfaction with that specific interaction at that specific touchpoint. An evaluation on the basis of a single interaction thus proves to be especially problematic in the management of actual employees. Additionally, the customer must once more take the time out to respond to individual questions or to a survey. Therefore, the question at hand is to what extent the customer sees an additional benefit of the survey in terms of relationship building with the company. Customer surveys regarding the NPS should thus be limited to their annual effectuation.

Moreover, a survey following every interaction also lacks consideration. Normally, a contact centre correspondent can sense the customer's level of satisfaction from the conversation itself. The former's incentive to record this information into a system for the logical development of customer relationships is, however, limited, particularly so in the case of problematic discussions. This is the kind of dilemma that can solved by the described systems. They actually measure true satisfaction, based on what the customer feels and experiences. This evaluation delves deeply into the customer's psyche at the moment the interaction is taking place.

By combining NPS, as a higher-level indicator, and the speech analysis touchpoint evaluations, it is possible to create an integrated control cockpit for customer service that not only allows conclusions to be drawn about the interaction quality and the customer's true experience, but that is also promotion-related. Unsatisfactory experiences are registered so that the customer receives specific treatment in the subsequent interaction in order to reestablish a positive experience. Retention campaigns can thus become even more targeted and logical thanks to speech analysis.

5.3.3 Chatbots and Conversational UI

Through an analysis of the spoken or written language, it can now be reflected upon how automated dialogs come about. The basis thereof is an infrastructure that has appeared on the smartphones of over two billion people since 2008 on "Messaging Apps" such as Facebook Messenger, WhatsApp, Amazon Echo, or the Chinese WeChat. Companies can now chat with their customers via this "Conversational UI". This has an advantage over the development of personal service apps, in that a generally accepted dialog infrastructure is used which is accessible and easily comprehensible for most users, and thus for customers (Sokolow 2016).

If an automation of the service dialog is to be reflected upon, simpler customer requests may be dealt with by chatbots, since more than 80% of questions posed in most industries are highly repetitive. The term chatbot is made up of two parts. The second part, "bot", is an abbreviation of the word "robot". This includes programs used for automatisation. The first part, "chat", refers to a specific function fulfilled by the bot in the communication mode. Therefore, a chatbot is a software capable of entering into meaningful dialog with people. The communication can be either written or spoken (Dole et al. 2015).

Chatbots are not a new invention. First applications were developed as early as the 1960s, at that time still being a completely programmed robot with a static reference framework. This is how the best-known case of "Eliza", in the role of psychiatrist, communicated with test subjects who felt convinced that they were conversing with a real person.

Recently, however, modern Chatbots have surpassed such "programmed machines" and are becoming increasingly developed. They must be taught through dialogs between customers and companies. In this context, one can once again clearly speak of Maschine Learning and the resulting AI (Iyler et al. 2016). In recent publications (e.g. Weidauer 2017), it becomes clear that in the increasingly precise conversation between bot and customer, the focus is not only on the system's learning speed but also on steering the customer through dialog using a skilful question technique. When the bot makes specific enquiries, the customer's decisions and, thus, his expression of will, become all the more clear. The precept "Who asks, leads" also applies to Chatbots.

Well-known examples of Chatbots equipped with AI and found in a real speech environment are Apple Siri, Google Now, Microsoft Cortana or Amazon Alexa (Sauter 2016). They accomplish nearly any personal assistant task. However, a chatbot can also be misused for automated reviews or other manipulation of public opinion (Sokolow 2016). In that sense, Iyer, Burgert and Kane point out that trust in new technologies, such as bots, is limited and should not be abused. As an example, they draw on the bot "Tay" from Microsoft, which used machine learning on Twitter in order to develop an

AI and "understand" how youth between the ages of 18 and 24 communicate. In doing so, the bot learned from the dialogs that were carried out with him. When the bot began making racist statements, as it had learnt in dialog, it was shut down and readjusted by Microsoft (Beuth 2016).

Such bots are newly integrated into the respective messenger environments and serve as conversation partners for the users or involve themselves in the dialog between several human users (Elsner 2016). The core idea behind this, is using the bot in order to automatically guide participants to products and services that play a role within the dialog. For example, an entire holiday plan, from flight booking to hotel reservations, up to the selection of excursions and restaurants, can take place in a single discussion, without navigating away from the messenger environment in order to look up commercial apps or websites that list prices and alternatives. Such transactions, which are concluded by means of communication are subsumed under the keyword "Conversational Commerce" (Sokolow 2016). If the chatbot is integrated into a popular messenger platform (Facebook Messenger, Slack, etc.), it simplifies the customer's day-to-day life, since less effort is required when, for instance, booking a flight via a short message rather than going through the entire process on the airline's application (cf. Annenko 2016). The bot's full potential, however, is realised only when one of its planned trips does not go according to plan: for instance, if the bot realises that there is a long delay on the flight as you are already on your way to the airport, it can autonomously make the booking changes required to ensure that planned reservation dates are respected. The customer remains unaware of this entire process and the airline is spared a profusion of unwished-for service dialog.

An example of chatbot use in customer service is the Digibank in India. It has implemented a chatbot that is capable of responding to customer queries and leading conversations in which the customer switches back and forth between different bank-related topics (Brewster 2016). The Bank of America adopts a similar position; here, too, customers can interact with a chatbot in Facebook Messenger. When taking a look at the Chinese Messenger service WeChat, one finds, for instance, that money transfers are carried out between chat participants and that all sorts of goods and services are ordered. In this case, it is easy to perceive the usefulness of the messenger environment as Conversation UI. In the case of a conventional product order, the corresponding eCommerce representational authority must be called over the Internet and the payment is generally performed either by a payment app or by a transfer within the customer's own eBanking environment. Just consider how many passwords a customer must enter within this set-up in order to authenticate themselves. An appropriately trained chat-bot can aid several customers rapidly and simultaneously, which is clearly a much more rational form of dialog automation.

Given that service requests occur in varying degrees of complexity, dialog monitoring has a particularly important role to play. This especially applies in cases where the bot receives new service requests. In such a case, the bot is unable to answer, or the answer is unsatisfactory for the inquiring customer. In the event that the bot does not "know anything more", it is essential that the dialog be taken over by a human correspondent. Subsequently, however, it is advised that the new service case be transferred back to the learning bot. In order to provide the bot with a basis of "service knowledge", Iyer, Burgert, and Kane recommend pilot testing bots with customers (2016). The risk of a bot that does not respond or that provides unsatisfactory responses should be reduced over time. Generally speaking, companies are only at the starting line of this development. Bots are slowly starting off by resolving highly standardised problems and gradually expanding into the complexity of dialogs (Simmet 2016).

5.3.4   Predictive Maintenance and the Avoidance of Service Issues

Predictive Maintenance is a mode of Predictive Modelling which is extremely important for the future of the service sector. Here, the treatment of big data and the Predictive Analysis that is based on it has a particular role to play, as a study by the University of Potsdam shows (Gronau et al. 2013). Predictive Modelling is characterised by, on the one hand, a high analytical degree of maturity, and on the other hand, by an increasingly high competitive advantage, emerging from the predicatively generated knowledge. As regards customer service, Predictive Maintenance primarily concerns a company's proactive behaviour to avoid foreseeable service issues (Hoong et al. 2013). It is therefore a matter of developing a model from available data sources, which predicts when a certain service issue could occur and what consequences it may have for the company and the client. If it is more convenient to provide the customer with a solution before the service event in question actually occurs, so-called irritants may be avoided for parties involved (Price and Jaffe 2008, also see Chapter 1 of this article). This is above all enabled by the fact that not only internal company data and information from the customer dialogs—as shown in Chapters 2 and 3 of this article—but also external environmental data, are used for modelling.

Hoong et al. demonstrate this with the use of a mechanical engineering example, as displayed in Fig. 5.12.

In contrast to a maintenance controller of this machine, which runs according to fixed times and usage cycles, the Predictive Maintenance model uses both internal and external as well as dynamic data to predict the machine's default probability. In purely economic terms, one can now consider the costs of a machine default on a daily or hourly basis. This is about optimising the maintenance or total maintenance costs. If maintenance is carried out prematurely, the equipment's wear parts could have been used for a longer period of time. This results in unnecessary costs. If the machine is incurred by the client company itself, the downtime costs can be passed on to the manufacturing company under certain contract conditions. Once again, machine learning comes into play. The algorithm learns from every machine default. Based on all running machines and their service intervals as well as unplanned defaults, the estimation model's accuracy continuously improves and can thus determine the optimal time for maintenance or replacement to occur.

This logic is also increasingly used in B2C environments for profitability purposes. For instance, consider the case of a trader who sells his customers high-quality coffee capsules under a club model at a high margin. Through its business model, this company knows its customer by name and address. On top of that it knows the amount and types of capsules that the customer has bought. At the same time, it knows the brand and type of machine used. The company is aware of the average life of this machine in relation to the water hardness degree at the customer's place of residence. In developed markets, this information is quite easy to find. The company also knows how often the customer has descaled his machine. The decalcification set is usually also obtained through the club. All of these factors result in an estimation model which is refined over time, as was described above. Now, it remains to see how the default "irritant" of this machine can be avoided. The retailer knows that a customer whose machine is down won't buy any coffee for about a month, until he has obtained a new machine. During this time, the risk is naturally higher, occurring in the form of supplier change, since a changeover barrier (a functioning coffee machine) has been removed. In order to minimise the lost sales margin and the risk of supplier change, the dealer now makes the customer an advantageous offer (from his point of view) as soon as the probability of a machine default has reached a certain level. The customer can (when ordering a certain amount of coffee) purchase a new (another one, from his point of view) coffee machine at what is for him an attractive price. If the customer accepts the offer, the Predictive Servicing has been successful for the coffee trader.

5.3.5 Conclusion: Developments in Customer Service Based on Big Data and AI

On the basis of the three areas of application presented, it can be observed that the use of big data and forms of AI, and therefore machine learning, is increasingly beneficial in the customer service world. With increasing advances in the fields of Voice Analytics and Predictive Servicing and the increasing dialogue capability of chatbots in a messenger environment, customers will be able to treat customer requests in a more automatic way, thus achieving cost and speed advantages. It remains to be seen how the challenges of machine learning and the selection of relevant data (value data) from the universe of "Big Data" can be overcome without losing customers through unsatisfactory dialogue on the way to automation. Along this path, the management of employees in customer service remains of particular interest. If it becomes apparent that person-to-person dialog no longer occurs, meaning that jobs are lost, it is questionable to what extent today's service professionals would actually teach the bots, thereby achieving the above-described efficiency gains.

5.4 Customer Engagement with Chatbots and Collaboration Bots: Methods, Chances and Risks of the Use of Bots in Service and Marketing

Dr. Thomas Wilde, University Munich LMU Media

5.4.1 Relevance and Potential of Bots for Customer Engagement

Obtaining information, flight check-ins or keeping a diary of one's own diet—all of this is possible in dialogue today. Customers can ask questions via Messenger or WhatsApp or initiate processes. This service is comfortable for the customer, available at all times via mobile and promises fast answers or smooth problem-solving. A meanwhile strongly increasing number of companies is already relying on this means of contact and the figures on chat usage speak in favour of this means supplementing or even replacing many apps and web offers in the future. The reasons for this are manifold.

Figures of the online magazine Business Insider[1] reveal a clear development away from the public post to the use of private messaging services such as Facebook Messenger or WhatsApp. Facebook meanwhile has a user base of around 1.7 billion people worldwide; 1.1 billion people use WhatsApp, and Twitter can nevertheless still record 310 million users around the globe. The platforms are growing fast, customers are accepting these platforms and are using them exceedingly intensively. And even technology has long grown out of the prototypes: IBM Watson won against a human being in the US game show "Jeopardy" in as early as 2011—the handling of customer dialogues in contrast seems to be downright simple.

5.4.2 Overview and Systemisation of Fields of Use

In principle, bots can be differentiated into chatbots and collaboration bots, depending on their area of use. Chatbots have direct exchanges with customers, prospective customers and other stakeholders and can be used in different places in marketing, sales and service, such as for qualifying issues in advance, providing leads with information (nurturing) or giving automated information in service.

Collaboration bots, on the other hand, support engagement teams in their work by proposing possible answers or routing options, taking over research tasks in knowledge databases or categorising activities and prioritising them dynamically.

The social media management provider BIG Social Media in its BIG CONNECT solution differentiates types of bots even further according to concrete application scenarios and makes available a suitable library of configurable bots (cf. Fig. 5.13).

Both chatbots and collaboration bots provide numerous advantages especially in marketing and service as they make 1:1 communication profitable where it was only given in exceptional cases in the past. In this way, entirely new services are becoming possible.

The chatbot as a virtual assistant can provide information about products and services in the scope of campaigns or customer enquiries, answer specific questions or take bookings/orders. The in the meantime significantly advanced process in natural language processing (NLP; Chapter 3) and artificial intelligence (AI; Chapter 3) make sure that the tasks bots reliably take over are becoming more and more complex.

On the use of spoken or written language, it must be noted that in view of usability, this is by no means and always the royal road. In fact, bots are to be considered as another user interface for a service that are to be created according to prevalent usability methods and give answers accordingly with list elements, graphics, etc., and are meant to use the various methods of input of the target platforms. Especially when using them on mobile end devices, it must be assumed that the communication partner has little interest in typing longer texts on the screen keyboard of their smartphone.

Collaboration bots, in contrast, are not used for direct customer contact but support the staff within the internal workflows. When humans process enquiries, bots can be used for intelligent routing, to search for information in the depths of the knowledge management system or for representing service cases. Their advantage is that they usually interact via simple interfaces with available software applications and can thus make use of numerous sources of data.

By using collaboration bots to optimise the handling of enquiries, around 50% of the costs can be saved, which would be incurred in Messenger or social media dialogue without such support, by routing, the preparation of proposed answers by bots as well as by bot-driven information research for a member of staff to answer an enquiry. If bots are used for the fully automated answering of enquiries, cost savings of a further 50% are possible in comparison with an intelligent solution in combination with an engagement platform operated by a member of staff, meaning that up to 90% less costs than with processing by telephone can be calculated.

5.4.3 Abilities and Stages of Development of Bots

Bots are, in fact, the big issue in the digital economy at present, yet they are not in principle a new issue: In 1966, Joseph Weizenbaum released the script-based bot ELIZA that allowed a person to communicate with a computer in natural language. When replying, the machine took on the role of a psychotherapist, worked on the basis of a structured dictionary and looked for keywords in the entered text. Even if this bot model only celebrated questionable success as a psychotherapist, such bots of the first generation with strictly predefined dialogue control and keyword controlled actions are still used in many places (Fig. 5.11).

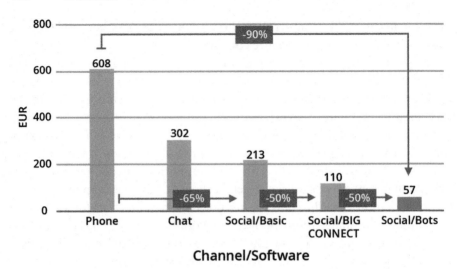

Fig. 5.11 Savings potential by digitalisation and automation in service

"Real" speech comprehension by way of NLP of a computer-linguistic methodology to be able to recognise and process correlations in meaning and contexts is nevertheless still rather seldom in today's practice, even though the processes have in the meantime reached market maturity. It is often the usability that puts a spoke in the wheels. Especially on mobile end devices, written/typed speech is not the means of choice for the convenient use of a service.

The second generation of bots that is primarily expected for 2017 does indeed continue to follow a rough process script, but already uses AI at crossroads. The question about the device used, for example, is "hardwired"; afterwards the dialogue partner can, however, send a photograph from which the bot can determine the device used including the serial number where applicable. Another example is the analysis of textual error descriptions. With the result of the analysis, a bot of the second generation chooses the suitable reaction from a list of given possible reactions and works itself through a "dialogue tree" with intelligent branching.

It is not until the third generation of bots where free dialogue and free conversation is allowed. This is made possible by the meanwhile wide availability of cloud-based solutions that not only provide scalable computing capacity for AI applications, but also skilled AI services as "AI-as-a-service".

5.4.4 Some Examples of Bots That Were Already Used at the End of 2016

1-800-flowers: The large American flower delivery service 1-800-flowers offers for the Facebook Messenger the possibility of ordering flower greetings per chat. The bot poses simple questions and then branches off the dialogue accordingly. Deliveries can also be made in German—however the destination must, however, still be in the USA at present. The set-up of the bot is simple, it recognises postal addresses from all over the world very reliably and offers a full selection and order process.

KAYAK: The travel portal provides a German bot for the search for hotels and flights. The set-up of the bot is equally very simple, it offers predefined choice options in the answers and thus sets the direction of the interaction. Deviations and free questions overchallenge the machine. The bot then asks the same question over and over as to whether you are looking for a hotel or a flight. No real dialogue can occur this way.

Jobmehappy: The chatbot of the job exchange Jobmehappy is equally simple but works reliably. The user asks a question that has to contain the term 'job' and a location or a job title. The bot immediately provides a choice of results—whereby a by all means meaningful AI was not made use of in this case either: Anybody looking for managing director positions will also come across "assistant to the managing director".

KLM: The bot of the Dutch airline KLM offers genuine customer service. Customers can change their seat, check-in via Facebook Messenger and receive information about their fight constantly. This way there is no hectic if the flight is delayed by a few minutes and the air passenger is still going through security. Once activated, the bot proactively informs the customer if the flight is delayed. The customer can ask the bot any questions around the clock—and what the machine cannot answer itself is evidently passed on to the service centre and answered from there.

This means that the bots that have been used to date still follow a clear, predefined script in dialogues. Most of them are nothing more than the reproduction of a search function in a chat application. Merely the KLM bot presented avails of a connection to the service centre and is capable of escalating service cases that the machine cannot process.

And it is precisely in this connection of the bot to the service processes and the available resources where there are, however, great potentials for customer service!

5.4.5 Proactive Engagement Through a Combination of Listening and Bots

The far-reaching possibilities the bots of the second and third generation offer are illustrated by the example of identification and use of customer engagement opportunities through social listening in combination with a suitable social media management solution.

Social media listening is, in the first instance usable, irrespective of the application of a bot. Social media listening in the classical sense (also called social media monitoring) describes the process where what is written and discussed on social media about a company, a product, a brand or an individual on the Internet is identified, analysed and evaluated.

Furthermore, active social listening is about providing information or offering proactive customer service faster—even before a customer requests it in the scope of a directly asked question. Active social listening thus enables companies to recognise business opportunities and to enter into a 1:1 dialogue before customers themselves seek contact—but possibly have with a competitor.

Two examples

A user posts a photograph of his car with lots of newly bought moving boxes on the car park of a DIY store. The text to go with it indicates an upcoming move. Through active social listening, a telecommunications provider, for example, can locate this specific post and posts a comment pointing out that the user should not forget to also apply in good time for their telephone connection to be moved.

The Facebook user "likes" the comment and contacts his telecommunications provider via direct message to ask whether he can apply for the move using this channel. Due to the fact that it has been noted in the CRM that the customer had asked for a faster DSL connection in the past, the customer is informed in the course of the dialogue on Facebook that an upgrade to the faster DSL tariff is possible at the new address without difficulty. The customer will appreciate this service—the offer was easy and quick for him to receive and the foresighted information about the faster DSL connection is the fulfilment of his request he had already expressed weeks or months ago. The entire dialogue can be conducted with almost no effort on the company's part as the process can be handled by a bot fully automated. The successful upselling has thus been realised extremely efficiently.

Another prospective customer asks about an electricity tariff and the associated incentives via a social media platform. The chat dialogue with the member of staff proceeds such that the user would like to switch, but he then remembers that he still as an energy supply contract that cannot be terminated until in a few months' time. The member of staff suggests setting up a reminder and the prospective customer agrees. A few months later, a bot contacts the prospective customer and points out that he would have to give notice now in order to then be able to switch over to the more favourable provider. If the prospective customer responds to this message in a positive way, the dialogue is immediately handed over to a member of staff who can conclude the contract. If the prospective customer responds and states a new time span again, he will be asked again whether a reminder should be set up. If he responds differently and names another provider, for example, indicates a move or other things, the bot can conduct the dialogue up to a certain point and then either hand it over to a personal adviser in the service centre or end the exchange. In all events, the customer feels valued as the company actively supported him. At the same time, expensive resources of the service centre are only put to use if the probability of a conclusion is high.

For these examples of use cases to lead to a sustainably positive customer experience for customers and to the aspired economic advantages for the company, appropriate technical solutions, an experienced project management department and structured process phases are needed.

Analysis phase: The basis of active social listening is a finely tuned monitoring of the social media platforms as well as the relevant concepts. The task in this early phase is to separate relevant from irrelevant comments and profiles. More flexible methods, in particular natural language processing and AI/deep learning, a simulation of the way the human brain learns, step in the shoes of simple keyword lists. Based on a social CRM database of past conversations and profile information, the attempt is made to identify similarities between the current data records on the social web and previous successfully developed leads, etc., and to classify them.

This way, a plethora of comments on the social web turns into "smart data"—i.e. data whose content and significance for the company can be clearly described and from which meaningful next steps can be derived. This is the way it is evaluated, for example, whether the user is classified as a "hot lead" or as a cold contact or whether a termination is to be feared. A flow of "engagement opportunities" is generated in this way.

Scoring phase: In the second step, the identified engagement opportunities are evaluated. Data from the social CRM allow for an inference of the prospects of success of an engagement and of the potential value of the contact. If a contact in the identified case is successful with a predictably high probability, the opportunity is given a high score and a contact is triggered with high probability. If the solution results in an interaction not leading to added value for the company (and/or for the customer), the interaction will not happen.

Next best activity phase: In the third phase, the decision is made as to how to proceed with a positively evaluated engagement opportunity. Its content, intention and value has been determined. The bandwidth of possible reactions is large: Users can be invited/requested to participate in campaigns, service offers can be submitted proactively, complaints can be anticipated and avoided in the best case scenario, leads can be generated and systematically developed.

Based on the collected and prepared information and in combination with data on active campaigns, a suitable activity is determined. This activity can either be executed automatically by bots or by a member of staff, such as in the service centre.

Execution/routing phase: The process is executed in the last phase. In the case of an automated process, a bot establishes contact in the next phase; in another case this is done by a member of staff in the service centre and they contact the customer. The knowledge gained from the first two phases can be used for the selection of the staff member (or the respective team), in order to route the service case to the right place based on skills. With regard to the first example mentioned, even the first comment under the customer's post could be made by the member of staff who avails of sufficient experience in moves. The system monitors this task being performed, records the implementation and controls that agreed service levels are adhered to. If the contact is performed fully automated, this is recorded in the system correspondingly—it is not until the intervention of a member of staff is necessary before the system in turn escalates it to the right team or team member based on skills.

Such models are applied, for example, today at Deutsche Telekom AG for the proactive processing of service cases as well as at Porsche AG for the early recognition and contacting of prospective customers.

5.4.6 Cooperation Between Man and Machine

When embedding bots in the process organisation and workflows in the company, three different models can be differentiated in principle:

Delegation: The bot takes over a process from the customer service agent.
Escalation: An agent takes over a process from a bot.
Autonomous dialogue: The bot activates itself according to predefined triggers and leads the user through the entire dialogue.

In the first model, that of delegation, the member of staff begins a dialogue with the customer, gives advice on products, for example, and passes the dialogue to a bot, who then carries out the subsequent booking. This takes the strain off the member of staff of entering a standardised booking and it automates the conclusion.

In the second model, the bot escalates a dialogue to a member of staff if the possible answers offered by the bot are not satisfactory for the customer. And if the customer wishes further advice, the bot passes the dialogue on to a member of staff. The example of the KLM bot described above matches precisely this model.

In the third model, the bot leads the user through the entire dialogue. A popular use case for this model are information services or the recording of fault reports. The triggers for the dialogue by the bot are incoming messages in a channel or the use of certain keywords.

The role bots take over in a dialogue depends in each case on the exact workflow in customer dialogue. The more AI is used in a bot and the more sophisticated its dialogue skills are, the greater the efforts in development. In most cases, 70% of all enquiries can be successfully automated with a simpler model—the 30% of the cases that cannot be solved by bots are then handled by a member of staff. In short: Which bot or which combination of bots is used depends on the profitability and the business model. In all cases, a bot supports the staff members effectively and takes the burden off them of repetitive tasks.

5.4.7 Planning and Rollout of Bots in Marketing and Customer Service

If the decision has been made to use a bot in customer dialogue, the dialogue has to be planned, the bot has to be developed and implemented. To this end, the objective of the automation and of the circle of recipients of the bot-driven dialogues has to be clarified. The bot's scope of functions is defined and then the dialogue structure to be reproduced by the bot is developed. In practice, a five-step process model has stood the test of time.

5.4.7.1 Step 1: Model Target Dialogue

Before automating a dialogue by using a bot, it is advisable to conduct this dialogue manually in the target channel for a whilst. The courses of the dialogue can then be coded and evaluated. The result is a precise overview of the typical courses of dialogue. All essential dialogue variations are recorded here. Afterwards, the decision can then be made as to which dialogue variations should be automated. The most appropriate wording and actions for every step of dialogue of the bot are then formed on this basis. At the same time, it is essential to identify the respective points where the bot enters and exits a dialogue. Dialogue paths that do not lead to a successful solution on a predefined path can nevertheless be successfully concluded later by passing the dialogue on to an agent.

It is also important in this first phase to already consider the aims of the bot and to define clear, measurable targets. This key data then provides the necessary control of success in the course of using the bot and, in turn, can be the basis for a granular adaption of the dialogues later on. Even the decision as to in which languages the bot is to be put to use is made in this phase.

5.4.7.2 Step 2: Integration into the Service Process

Chatbots can be integrated into the service process in three ways (or a combination thereof) as previously described:

Delegation: The bot takes over a process from the customer service agent.
Escalation: An agent takes over a process from a bot.
Autonomous dialogue: The bot activates itself according to predefined triggers and leads the user through the entire dialogue.

The decision in favour of one or several use cases defines whether the target dialogue will be fully or semi-automated. It also defines when and how the bot becomes active or inactive. This means, for example, that the bot is activated by an enquiry in Facebook Messenger, greets the customer, asks him about various search parameters in a dialogue and then plays out search results in the shape of a list of links. After that, the customer is either bid farewell or asked whether he wishes to carry out another search. This could equally be handled by the bot. Points of escalation in this concrete case can be set at asking for the search parameters and when saying goodbye. In both cases, the customer could be passed on to a staff member at this point, who then takes over the dialogue. Equally to be defined during the process integration is which groups of agents may pass dialogues to the bot or to which groups of agents the bot may escalate dialogue.

5.4.7.3 Step 3: Choice of Software and Bot Configuration

After all basic points have been clarified in Step 2, the best software solution has to be chosen as well as the course of dialogue, activation criteria and abort criteria have to be recorded in the configuration. After the fact that mostly delegation or escalation occur in practice, the focus in this phase lies not only on the choice of the technologically suitable bots; the bot's environment must also be considered.

With that, it is important for the software to support all target channels and have flexible configuration options for the dialogue, smooth routing between the bots and staff members as well as monitoring, intervention and reporting functions. Solutions for customer engagement bring along extensive libraries of pre-configured bots that can each then be adapted to the intended use.

5.4.7.4 Step 4: Bot Testing and Deployment

Before the bot is actually put to use, it has to be tested internally. All dialogue steps must be documented precisely and the reporting must provide the results of the test that were previously laid down in the definition of the key data.

If the bot works as planned, live operation can be started on the various channels. Upon deployment, it is again a matter of setting the correct conditions for the activation. Should public or private messages on social media be responded to? Which keywords must be contained in the enquiry? Should public enquiries also be answered publicly or preferably privately?

It is possible that there needs to be a test in live operation as to whether the referrers have been transferred properly. This is always important when the bot refers to a website. This way, it can be traced later on how many accesses to the website were generated by the bot. Even the multilingualism has to be tested once again and may also have impacts on the escalation—the staff member taking over a dialogue from a bot must be able to continue it in the chosen language.

5.4.7.5 Step 5: Monitoring, Intervention and Optimisation

It is advisable to first fully and later randomly monitor the dialogue quality of the bot. If necessary, individual dialogues can be taken over and assigned to an agent.

Furthermore, it is important to search for signs of usability problems in reporting:

What is the percentage of dialogues that went according to plan?
How high is the percentage of successful endings or how high is the escalation quota?
How high is the percentage of dialogues aborted by users?

This key data gives a good overview of whether the user gets along with the bot. At the beginning, the figures can be measured on the results of the internal test phase; in the course of time a more accurate picture of the acceptance and efficiency of the bot is given.

5.4.8 Factors of Success for the Introduction of Bots

If we consider the successful and unsuccessful bot trials, the "race" for the use of bots in customer dialogue that has already commenced can no longer be followed blindly. Most recently, Microsoft made the headlines with the Twitter bot Tay, which was originally meant to be proof of performance of modern AI skills. Within one day, the bot learned a lot from his contacts on Twitter and turned from being a youthful chum to a "hate bot …, who uttered anti-feminist, racist and inciting Tweets".[2] Such a loss of control over a bot would have severe consequences in a company's customer service.

When planning and implementing bot projects, the following points must be considered.

5.4.9 Usability and Ability to Automate

Many service cases require human intelligence and empathy. These cannot be replaced by a bot—at least not in an economically meaningful context. All in all, however, a large number of service cases can be identified that can be automated by the use of bots. Bots are always unbeatable when it is about reproducing a clearly defined dialogue path for the user.

Practice shows that users tend to avoid written communications especially where the communication is frequent and when communicating from mobile end devices. In these cases, efficient and standardised communication, which is to be supported by a suitable user interface (with bots the likes of platform-specific input options), is desired. Moreover, it can also be ascertained that dialogues with clearly predefined answer options get the user to the end more quickly in the majority of cases.

5.4.10   Monitoring and Intervention

The Microsoft example demonstrates that bots have to be monitored. This applies not only to self-learning AI bots but also to the dialogue of simple bots. User behaviour sheds light on where the bot can be optimised. This further development and optimisation subsequently leads to better customer experience.

In customer service, bots need human partners that can always jump in when the bot considers cases incomprehensible. This interplay between man and machine takes the strain off service agents with repetitive and trivial enquiries and creates room for empathetic customer dialogues and high-quality service.

In order to enable this, bots must be connected to existing service processes. The productive cooperation between man and machine must then be orchestrated by a software solution to avoid interface problems.

5.4.11   Brand and Target Group

Does the use of a chat bot match the brand and the source of communication? And does the target group want to use Facebook Messenger or WhatsApp for such communication? Even if the use of bots has been accepted in the customer service of today, the scenario has to be adapted precisely to the brand and the target group. In the close analysis of the services and customer behaviour, it may, for example, turn out that a chat bot is not necessarily the first choice for addressing silver surfers, but a collaboration bot can prepare a large number of enquiries in service for the staff members. Generation Y, in contrast, may quickly turn its back on a trendy brand if they are not served quickly and efficiently on the channels they are used to in everyday communication. This is why the analysis prior to the actual application is of such great significance.

5.4.12   Conclusion

The saying "service is the new marketing" has been accompanying us for years—with bots, there is now an economically attractive way to actually develop this to become a substantial and bearable pillar in the marketing mix. This is how a bridge is built between service organisations that are eager for avoiding and limiting contact, and marketing organisations that invest a lot of time and money in establishing and continuing partly the same contacts. This way, issues from marketing, sales and service ideally become focuses of a customer communication that is indeed paused every now and again, but can always be picked up on.

For the use of bots to improve efficiency, quality and reaction time in service and for the situation-related dialogue in marketing to be designed in a sustainably successful way, there needs to be a workflow and engagement solution that controls the agent and engagement team dialogues and bot dialogues in equal measure and enables a simple handover between man and machine. Social media management solutions provide an excellent starting point for this.

5.5   The Bot Revolution Is Changing Content Marketing—Algorithms and AI for Generating and Distributing Content

Guest contribution by Klaus Eck, d.Tales GmbH

The subject of AI has become increasingly popular in companies ever since the beginning of 2017. It is co-responsible for the search results on Google or Bing. In addition, some of our digital assistants on our smartphone as well as some messenger bots are based on (simple) AI.

At the end of 2015, Google extended its algorithm by AI: Google RankBrain. Behind it is a system that learns little by little more about the semantics of user queries and which increasingly improves with this knowledge. The aim: RankBrain is meant to fulfil the users' needs in an increasingly better way. And with it, Google has taken the first step towards self-learning algorithms. Many upgrades will be possible in the future without any human assistance, because the systems will learn something new all on their own.

AI will also play a significant role in content marketing when it comes to combining contents with each other and promoting them. What still sounds like dreams of the future will be totally normal in a few years. The abilities of artificial AI are said to go to such lengths that it can automatically publish and distribute content on various platforms.

AI already offers useful features for companies that would like to operate on an international basis. With the help of algorithms, Facebook is able to translate a post into the user's respective mother tongue. This depends on the given location, the preferred language and the language in which the user normally writes posts. The cumbersome multi-posting of contributions can thus be avoided.

AI is used for, among other things, optimising the targeting of adverts and search engines. As well as that, information can be tailored to the users' needs more efficiently in the bot economy.

5.5.1   Robot Journalism Is Becoming Creative

Algorithms are able to automatically search the Web for information, pool it and create a readable piece of writing. In addition, data-based reports in the area of sport, the weather or finances are already frequently created automatically today.

Recently, for example, merely a few minutes after Apple had announced their latest quarterly figures, there was a report by the news agency Associated Press (AP): "Apple tops Street 1Q forecasts". The financial report deals solely with the mere financial figures, without any human assistance whatsoever. Yet, AP was able to publish their report entirely via AI in line with the AP guidelines. For this purpose, AP launched their corresponding platform Wordsmith at the beginning of 2016, which automatically creates more than 3000 of such financial reports every quarter, and which are published fast and accurately. It is no longer that easy to distinguish between whether an algorithm or a human has written a text.

Another exception of recent times is represented by the IBM invention called "Watson": After its victory in the quiz show "Jeopardy", Watson showed what is already possible with AI in the field of robot journalism. As the editor-in-chief, Watson created an entire edition of the British marketing magazine "The Drum". Thousands of copies of the edition were printed, in which he had both selected images, adapted texts and designed the pages. Creative AI that—as was to be shown in the test—works excellently.

To this end, he was fed with data about the winners of the "Golden Lion" from the Cannes Lions International Festival of Creativity. It was not only about creating the magazine, but at the same time, about creating AI that suited the taste of the lifestyle public. Watson was thus meant to create something that many brands have not succeeded in doing to this day: Place the stakeholders in the spotlight and align the content marketing activities with their interests and needs.

5.5.2 More Relevance in Content Marketing Through AI

It would be conceivable, among others, for AI to adapt texts that have already been created to the linguistic habits of different target groups, so that a medical text, for example, could be understood by both doctors and ordinary people by having medical terms explained.

It is merely a question of time until algorithms are able to write texts for any target group whatsoever. In the future, AI will presumably even be able to produce excellent content at an enormous speed. This way, texts can be individualised and personalised more easily so that all essential information is included via a reader and which affects the written and adapted text.

AI becomes very familiar with the readership in this process and can utilise all information about the recipient in such a way that every single piece of content is unique. Just imagine the content that would be produced if AI could read out your entire (public) Facebook profile and were able to use this information for matching content.

In principle, it would suffice if retargeting were not used for advertising but used for the targeted addressing of content. In content marketing, algorithms are more and more frequently taking over this task, which is necessary for the targeted play out of the content as well. In addition, contents are played out in an appropriate context (content recommendations). Instead of one article for all, personalised content will be possible on the basis of AI, and which are closely based on the reader's respective range of interests. The result of this is unique contents in the logic of mass customisation because the AI knows their readers and responds in a personalised way. Everyone receives their own personal content.

5.5.3 Is a Journalist's Job Disappearing?

The fear here is that the journalist's job is disappearing completely. However, AI can also be very helpful in journalism. That should become apparent especially in investigative journalism. Algorithms can help in linking similar information and in extracting individual specifics from general data. The task at hand is to be able to recognise patterns and hypothesise.

This is where big data and AI intertwine when, for example, extensive data has to be studied and correlations have to be found. Journalists could then leave the analysis part, which takes up an awful lot of time, to AI and then fully concentrate on writing their article.

The point is to implement AI at the right places in a profitable way, not to simply replace the journalist. In addition, AI systems first have to learn ethical standards. This was demonstrated, for example, by the Microsoft bot Tay, who was meant to simulate a typical American male or female youth and communicate directly with the users on Twitter: He had to be switched offline in no time at all because a lot of users taught him racist content. It thus becomes apparent that even bots require some kind of guideline. Bots also have to observe certain standards in the same way a journalist has to stick to editorial guidelines.

AI is an exciting development for content marketers and will make a huge difference to the job profile in the future. After all, they are being given a tool with which content creation and distribution can be automated in many areas at a high standard of quality. Even now, there are endless numbers of posts on the Internet that have been produced and published by algorithms.

In the years to come, we will get to know many examples that will make it obvious how much mass-customised content will stand out from general content. Anybody who feels personally addressed mostly also reacts in a positive way. There will be hardly any way of avoiding a corresponding personalisation of content marketing. This will have effects on the role of and demand for content creators (journalists, writers, etc.) but all in all, will promote content marketing.

First of all, we will get to know AI via bots in everyday situations, which will be able to respond to individual enquiries via messengers. They can provide the customer with directly individualised content by extracting the information needed from the database in a split second. This way, every customer receives information customised directly to their questions and needs.

Bots can equally make the information available on platforms that is relevant for every single customer, meaning that, in combination with the corresponding algorithm, not a general but a customised news page can be created, which is adapted to each individual user in their current situation.

5.5.4 The Messengers Take Over the Content

A few billion people have already moved their communication from the World Wide Web to the messenger world of WhatsApp, Facebook Messenger, Snapchat and WeChat, etc. The people online are thus leaving the digital public domain and are now difficult for brands to reach. They are moving around in the part of the digital world (dark social) that is "invisible" to others, are no longer sharing their content with everybody via their newsfeed on Facebook, for example, but are restricting themselves to sharing their content by messenger with a manageable circle of friends.

This will change a lot in comparison with apps and browsers. The interface is focused on chatting with real people or with bots. There are messaging apps, chat bots and voice assistants. Users can, for example, use their voice to ask Siri or Google Assistant on their smart phone about the current weather, turn on the light using a voice command via Alexa, play a piece of music or have the news read to you. WeChat offers even more possibilities that are used by more than 800 million people worldwide. Via the Chinese messaging app, invoices can be paid, services can be ordered and even payments can be made to friends.

5.5.5 The Bot Revolution Has Announced Itself

There has been a huge hype about chat bots since 2016. Every content officer should take it very seriously. These bots can have a radical effect on content marketing. If we were to receive all contents via an interface like Facebook Messenger, WeChat or Telegram that was previously only available via browsers, newsletters and apps, a thrilling alternative for content distribution would be formed. After all, bots can provide us with relevant information in the right context in the future. Ideally, we will receive less and better information this way and thus avert the content shock.

Most bots are simply answering machines that are similar to a living FAQ list or a newsletter. Only a few of them count among the league of AI. If we ask a bot a closed question, we will first receive a simple answer without any surprises. In most cases, bots cannot respond to spontaneous human behaviour and open questions. Instead, we receive a counter question along the lines of: "I don't know what you mean by that". Bots are far from human empathy. In most cases, they only give pre-worded answers on the basis of a database in which all possible responses are listed. This is the reason why bots try to direct us in a predefined direction that they can understand again. Unexpected courses of conversation lead to the end of them. Furthermore, the AI-less bots can only respond to standard questions, do not remember and do not really learn anything new.

Yet, that is not meant to beguile of the innovations expected in the future. In combination with AI, bots become powerful tools and self-learning systems that understand our questions better in the course of time and thus give us the right answers, because they understand our context. Virtual assistants that have comprehensive access to our personal data are able to give good answers because of this database, which saves us having to search and sort out knowledge. This is where the actual bot revolution, which is announcing itself with gentle steps with the simplest of functionalities, lies. Many brands are already preparing themselves for this development.

The bot revolution is changing the way and means of how brands obtain access to potential customers via their supply of content. By 2027, this development will have great consequences for the marketing and communication world and will radically change previous communication models. Conventional models that rely on one brand message for all will function less frequently.

Content strategists will have to develop a certain feeling for the fine changes in the communication and content mix so that their organisation can react in good time to the changes in the digital continuum. After all, they do also want to reach their target groups with the brand messages in the future.

5.5.6 A Huge Amount of Content Will Be Produced

Having more content on one's own website can no longer be an adequate solution in the future. The race to get the first of Google's rankings will become cumbersome if fewer and fewer people take the route there. Due to the changes in the Google index, the search engine optimisers have long been relying on the quality of content in their SEO measures.

Brands should not feel secure when it comes to positive feedback on their content because, according to the study Meaningful Brands by the Havas Group, in 2017, 60% of the customers worldwide do not regard the content produced by brands as relevant. That is not a good outcome for the group's activities. Good and meaningful contents do, however, have a positive impact on the market success of the brands. After all, 84% of those questioned expect brands to produce content. It is thus worthwhile to pay attention to the quality of content.

In content marketing, marketers measure quality on the basis of the results they have achieved with the content. There can be completely different targets: For example, reputation, leads or engagement.

The content is relevant if it fulfils the stakeholders' needs and addresses them emotionally. If the topic of a brand or person receives no feedback whatsoever, this is because too little consideration was taken of the benefit for possible readers. It all depends on the packaging of the ideas if brands wish to get through to their customers with their topics.

When too much good content is being offered, meaning that nobody can or wants to filter the many results for themselves any longer, we then speak of content shock. At some point in time, nobody can or wants to perceive the wide range of content. That started to become a problem at a very early stage with the development of the Internet.

The World Wide Web began its triumphal procession at the beginning of the 1980s. On 6 August 1991, the computer scientist and physicist Tim Berners-Lee, who was employed by the Geneva CERN, presented the World Wide Web project to the public for the very first time. Two years later, CERN made the Web freely available to the world. The breakthrough for the general public was from 1993 onwards with the first browsers Mosaic and Netscape. Tim Berners-Lee was not necessarily a fan of the browser idea. In 1995, as the Director of the World Wide Web consortiums, he voiced criticism about the browser concept: "There won't be any browsers left in five years' time at the latest". The world of browsers outlived his forecast by many decades, which does not mean that we will still be living in a world in which the Internet is dominated by browsers in ten years' time. The current development of the platform Google, Facebook, Amazon and Snapchat rather points towards the opposite.

5.5.7 Brands Have to Offer Their Content on the Platforms

Several billion people prefer the messenger on Facebook, Snapchat, Telegram and WeChat. They are thus no longer moving around in a public, web-based world but on their platforms among their own, consuming content and communicating with each other there. Previous content activities will thus be radically put to test in the next few years. Anyone who wants to continue to reach their stakeholders will have to make smart content offers that depict their needs and provide them on the preferred channel. When a brand offers a lot of content but only on their own content hub, at first glance, it appears confusing because orientation does not always come easy.

Three quarters of companies rely for the main part on owned media, in order to distribute their brand messages via it. In 2027, many will shake their heads at this misconception. On their own websites and social media channels, companies do not necessarily reach potential customers, but rather those who are already in contact with the brand anyway. If a company wishes to reach new customers, they should preferably go to where they actually are. This is the result the study "Content Marketing and Content

Promotion in the DACH Region" the online marketer Ligatus came to. It is much easier to transfer the content to the adequate platforms, to publish it directly in close digital proximity of the stakeholders than to urge them via shares and SEO/SEA to change over to our owned media. Therefore, bid farewell to your own website and preferably rely on a well thought-through content distribution.

5.5.8   Platforms Are Replacing the Free Internet

People online love social media and prefer to stay on the respective platforms, according to the international Adobe Digital Insights (ADI) EMEA Best of the Best 2015 Report that was presented in July 2016. It is difficult to lure them away. Links on some of these platforms are simply annoying and will hardly be of any importance for winning traffic in 2027. In times of Instant Articles, LinkedIn Pulse, Xing Klartext and Facebook Notes, social media users spend time in their networks. On WeChat, the users can obtain their entire content via special WeChat pages directly in messenger. Nobody needs to use the browser for this anymore.

Even the Germans stay in the social web and prefer digital cocooning. They hardly ever visit external websites from there. The average website traffic rate out of social media in Germany is no more than 0.54%. More than 99% thus remain on social media without visiting websites from there.

Only a few companies have reacted to this so far. They are still placing their focus on content marketing and less on content distribution. In order to achieve an adequate reach with high-quality content, increasingly more companies will have to supplement their contents via content promotion on other portals.

5.5.9   Forget Apps—The Bots Are Coming!

In 2007, the triumphal procession of the smart phone began with the introduction of the IPhone by Apple and with that, the era of the apps. This completely changed the way people interacted with the world. There seems to be an app for every kind of problem. Thanks to the direct mobile access to information, many business fields have changed radically: From customer service over marketing down to communication. Many companies are relying on their own apps which, unfortunately, are being accepted less frequently.

5.5.10 Competition Around the User's Attention Is High

It is difficult to be successful in the app stores because not all of us actively use that many apps on our smartphone. At present, more than 50,000 new apps are offered every month in the Apple Store alone. Of those, only a fraction are downloaded and even fewer are used. At the end of 2015, for example, there were around 1.5 million apps in the Apple Store that recorded hardly any or no downloads anymore, resulting in the app store being called was even referred to as the app cemetery on Techcrunch. Most users spend their time with five apps only. A download alone does not suffice to be successful. Only three percentage of the apps are still used after 30 days. 65% of the users do without the download of further apps. In comparison with that, three billion people use their messenger 17 times a day.

5.5.11 Bots Are Replacing Apps in Many Ways

Companies should prepare themselves for the farewell to apps. Their place will be taken by chat bots that will assume many of the apps' tasks without there being a need for downloading a new app. Due to the direct delivery of content via messenger, apps could more quickly become less important in the future. Since the launch of the Facebook Messenger platform in the spring of 2016, more than 34,000 bots were launched there by the beginning of March 2017.

The greatest change in the bot world is that we obtain all applications via an interface like Facebook Messenger, WeChat or Telegram that used to be distributed among various apps. In the future, bots will thus provide us with all information we need in our everyday life. Voice and text will serve as the user interface. Bots can substitute search engines as well as replace websites and shops. And in addition, bots alleviate making appointments, play music and assist us in making payments and in communication.

5.5.12 Companies and Customers Will Face Each Other in the Messenger in the Future

The app and web world will increasingly become a data-based bot economy in which we will come to appreciate voice and messenger as new content hubs. We will receive out context-based contents via these so that they are more relevant to us and available faster. Ideally, there is less, but the right content instead.

The messengers will be the first touchpoint where the interaction with the customer takes place. According to the renowned "Mary Meeker's Report", messaging has gained greater significance for millennials in communication than social media. If brands wish to reach the young target group, they will have to further develop their range of information and dialogue accordingly. In a few years' time, messenger will make other customer communication channels seem less important. Neither apps nor other social media channels will demonstrate comparable significance.

5.5.13 How Bots Change Content Marketing

When considering the future of content marketing, one aspect is of particular significance that nobody who wishes to be successful in the long run should neglect: AI and bots will become game changers in a few years. Many of the former content strategies will be turned upside down by the new possibilities and thus become a greater challenge to companies. Some experts thus speak of the death of the (former) content marketing by the AI algorithms. This is certainly an exaggeration, even if provided with a spark of truth.

Content marketing itself is regarded as one of the most cost-effective marketing strategies that is asserting itself increasingly more worldwide. Even if it is not always easy to be visible on the Internet with one's own content, one thing remains certain: Customers have a great need for information and want to be entertained. Despite the content shock, the best and most unique contents will always assert themselves somehow. If the demands on content change, then brands and media have to react by responding to this, presenting their contents more visually and changing the channel they are played out on if necessary. As long as content marketing quickly reacts to the stakeholders' interests, it is successful for the main part.

Due to its usership on several platforms, Facebook has sufficient data to be able to analyse the way communication takes place on digital channels. Anybody who best understands how their customers communicate could use this profound knowledge for the set-up of their bots, and give them much more AI. Whilst bot providers have to understand what the messenger users want, people learn at the same time how best to speak with bots. The expectations of bots, however, have quickly decreased since the beginning of the hype around the virtual assistants. Most bots are too rudimentary. They frequently only appear as small FAQ assistants that can only respond to a few questions. Yet, this could change very quickly with the slowly growing number of AI-based bots.

5.5.14 Examples of News Bots

The US news channel CNN is one of the first to appear with a bot as news provider. CNN offers very much in comparison with other bots. The bot learns which topics the listeners like and personalises the news very well. Via this channel, we can receive regular content about our desired political affairs.

In comparison with that, the **Novi Bot** seems to be very simple: The young media offering Funk from ARD and ZDF news offers its bot in the style of a chat platform. The news bot delivers a compact news summary twice a day and is thus meant to address mainly the 14- to 29-year olds via Facebook Messenger. At the same time, the Facebook Messenger texts are supplemented by short videos, GIFs and photographs. They each make reference with a link to the own background reports.

When starting up Messenger, the user learns in an offhand style: "I'll be in touch twice a day – with the news that is the most thrilling! Short and sweet in the morning, more detailed in the evening. (You can unsubscribe at any time by writing 'push'.) Use the buttons below to read the news. Sorry if there is a hitch every now and again – I'm still a bit beta". The news is teasered cheekily and it is linked to the respective online news channels of ARD and ZDF. **DoNotPay** is a successful bot lawyer that began with specialising in fines for parking violations. The chat bot of the Stanford student Joshua Browder automatically checks to see whether the fine can be circumvented. Until March 2017, he was successful with 64% of his submissions to the authorities and, by doing so, he has saved around 160,000 users a total of four million in paying fines. Such a robot lawyer is also imaginable in the case of flight and train delays.

At the beginning of 2017, DoNotPay added a further means of support in administrative formalities: Until April 2017, the company offered asylum seekers in the countries of Canada, Great Britain and the US assistance in applying for asylum and helped to avoid making formal mistakes. In this way, Browder wants to help people who are fleeing, who cannot afford a lawyer. Foreign-language asylum speakers in particular are to be helped in understanding the complicated immigration forms. To this end, the chat bot asks some questions of the person seeking help on Facebook Messenger that help them with filling in the forms. **Travel agency:** There are the first information and booking offers in tourism via which travellers can plan their holiday. In the meantime, the bot offers for Facebook Messenger and WhatsApp are rapidly increasing. Instead of going to a website, tourists can get information from personal bot travel agents by asking their questions directly. They receive their answers automatically without any waiting time whatsoever. Among the first offers are the flight search engine Skyscanner, the meta search engine Kayak, some airports and airlines such as Lufthansa as well as the tourism portals Booking and Tripadvisor. In many cases, the bot offers appear to still be very rudimentary. They can only manage complex enquiries in rare cases. The Skyscanner bot, for example, can find a flight to New York but not go into detail about a specific fare. The booking interfaces of travel websites are still clearly superior to it. Yet, according to the opinion of some bot providers, this should change in as few as a couple of years.

With Lufthansa, airline passenger can establish direct contact via a bot and look for the "Lufthansa Best Price". The bot Mildred is then meant to find the cheapest outbound flight within the next ninth months plus the return flight, all in a split second. The booking itself is then conducted directly at lufthansa.com.

In e-Commerce there are numerous examples of successful bots. In November, Nexxus launched the bot Hair Concierge which, with the help of AI, answers questions on hair problems and makes direct reference to individual products so that the customers can purchase them directly via Facebook Messenger. In January 2017 alone, Hair Concierge received more than 450,000 messages. For the bot promotion, Nexxus mainly relied on influencers, word-of-mouth advertising and social media sharing at the beginning. This way, the bot managed an enormous organic reach without any use of paid media whatsoever. Bots are becoming particularly important in call centres. This is shown by an example of the telecommunications enterprise Vodafone, among others. Their virtual agent Hani answers about 80,000 enquiries per month and thus replaces some call centre agents. After all, he can answer 75% of the questions.

5.5.15 Acceptance of Chat Bots Is Still Controversial

Nobody knows exactly how messenger users will react to chat bots in the future and whether they will engage in using and communicating with bots. The results of the W3B Report "Trends in User Behaviour" from the beginning of 2017 show that many people online are skeptical about the new tools. Only a few of those questioned can imagine a usage for the dialog on websites or shops. At present, three quarters of German online shoppers prefer online communication to be by e-mail or online form with real persons of contact. Whilst every fifth online customer can imagine establishing contact with a website or shop operator via chat, only four percentage of online shoppers want to communicate with a bot.

Twenty eight percent of those asked accept chat bots in principle. In contrast, 50% of online shoppers object to them, mainly because they find the means of communication too impersonal. That is the key argument for 60% of the objectors. Many find the technology is not yet mature or see absolutely no benefit in the bots.

The concept of the chat bot is, however, still a very new one. There are huge differences in quality among the respective bots on Facebook Messenger, making the evaluation of all bots in one survey very difficult and which actually says rather little about the actual social acceptance.

"Online users today are still very critical of the use of chat bots in customer communication in comparison with other technological trends such as Smart Home or VR2", says Susanne Fittkau, managing director of Fittkau & Maaß Consulting.

Other studies see a greater acceptance of the bot. A survey by the digital association shows that every fourth German can imagine using bots. International studies give even more reason for hope. According to the analysts from Mindshare, 63% of those questioned can imagine communicating with a company or a brand via a chat bot. As a rule, users can expect very fast and good answers. If the expectations are not met, the chat bot experiment is too expensive for a brand. 73% of all Americans would not give a bot a second chance.

The man-machine dialog is completely unfamiliar for online people and is still completely at its beginnings. Even the first experiences with Siri, Google Assistant, and Cortana, etc. do not suffice for this. Anybody setting up a bot should thus attach great importance to accompanying communication for the offered bot so that the benefit is explained. Even content marketing for the bot can by all means be recommendable to introduce these innovative technologies to customers.

For marketers, bots are tempting because they promise access to billions of people on messengers. Due to their very simple interface up to now, bots are currently best suited for simple, direct questions. Bots will not be able to become a real alternative to apps and websites until the bot makers succeed in making the customer dialog a recurrent experience with their provision of information.

Bots are promising in customer service because they can improve the customer experience by 24/7 access to important simple information despite automation. In comparison with a call centre agent, bots respond in real time, are always available and always stay friendly. As robots, they do not know stress and thus appear to be very pleasant.

An intelligent chat bot should be just as good as a call centre agent. Due to a good connection to AI, which can expand the skills enormously, bots can learn from their customer experiences and optimise themselves independently. This way, customer relations can be improved on the whole.

The analysts of Gartner thus even anticipate that by 2020, around 85% of all customer interactions will do without human customer service. This, however, presupposes a good data base and a fundamental knowledge of customer enquiries. The better I cover all customer needs with my bot contents, the more likely the acceptance of chat bots in society will increase.

Many robots are intentionally shaped with childlike characteristics that we find likeable. For this reason, an independent digital personality is important to quell our fears about dealing with bots. We do want to know whether we are communicating with a person or a bot. At the same time, bots should have an intelligent and personal impact on us as is possible so that we are able to trust our virtual assistant.

Due to the use of bots, in combination with AI, users obtain communication and content offers customised to the respective needs. This reduces the complexity of the multifaceted contents on the Web. Bots who select the right and thus relevant content for us, assist us in arriving at our results faster. They substitute the cumbersome online search and replace previous web and app offers.

The better that is attuned to our needs, the sooner the acceptance of bots will increase. In addition, nobody should forget that bots can be the face of a brand, comparable with salespersons, an advertisement or even individual websites. After all, they convey the very first impression of a brand and should match the other customer experience. Customers forgive inconsistencies rather rarely.

5.5.16 Alexa and Google Assistant: Voice Content Will Assert Itself

Many people seem to have got used to voice very fast. We receive our content via Siri, Amazon Echo and Google Home on demand. Since 2017, voice control tools in the context of smart homes have been the stars of many technology exhibitions. The acoustic recognition of human speech changes the way in which we can access information. Instead of typing and pressing a touch screen, commands are simply spoken.

This way, we direct our questions to the virtual assistant at Amazon Echo and receive really well-spoken answers even today. It is the easiest means of access conceivable: The spoken human word. The speech recognition via

Alexa and her relatives will accompany us everywhere from childhood in a few years: Barbie manufacturer Mattel already offers a digital babysitter with speech recognition. Numerous car manufacturers are extending their digital range with virtual assistants. Amazon itself not only relies on Echo and Echo Dot, but also introduced numerous Alexa-driven products at the CES 2017. These included cars, TV sets, other loudspeakers and refrigerators.

No surprise then that market researchers from Gartner see a huge growth market in voice-controlled devices. Voice and bots will fundamentally change or replace search behaviour in a few years. Mere keywords that we currently enter into search engines will become complete questions, answered by bots. Instead of typing, more and more people are relying on their voices and the help of voice assistants such as Siri, Google Assistant or Cortana. This will further affect the way in which we search. At the developer conference Google I/O 2016, Google CEO Sundar Pichai revealed that currently 20% of all searches are conducted by voice search. Most of them are used for calls, telling the time, the current cinema program or for navigation. 50% of all search queries are to be conducted via voice search by 2020.

The search engine group introduced Google Assistant and Google Home to the market at the end of 2016. Networked homes can be controlled by these. Amazon offers a foretaste of the future Google scenario in the USA. Nine million homes there use Amazon Echo and Alexa on a daily basis and are getting used to posing questions using their voices. Instead of typing and searching for content in a browser or placing orders there, the interaction with brands is taking place with the voice as the interface. This means browsers could become increasingly superfluous in the future.

The analysts of Gartner are expecting 2.1 billion US dollars for the new interactive loudspeakers by 2020. These digital assistants should then be distributed with their hardware in about 3.3% of all homes worldwide. Amazon Echo, etc., can respond directly to voice commands and play a film on the TV, read you an e-book, turn off the light, play music on Spotify or find a train route.

5.5.17 Content Marketing Always Has to Align with Something New

The future of the success of content lies in questioning the familiar. The touchpoints are changing faster than many a marketer expects due to technological innovations. At present, everything is concentrated on the browser world, because many people obtain their information in this way. Yet, the information overload is increasingly overstraining people online, who thus prefer to take "short cuts in Digitaly" and distance themselves from links, because they usually are perceiving enough information around themselves.

5.5.18 Content Marketing Officers Should Thus Today Prepare Themselves for a World in Which ...

- The website still does exist as a content hub, but now only leads a digital existence in the shadows,
- We only accept excellent content that is played directly off the platform itself,
- Spoken content takes up more space,
- Multimedial product information replaces classic text style,
- The content in our physical world is also present digitally (augmented reality),
- Inbound marketing is the only functioning marketing method, and
- Texts may still be justifiable as script, but the visualisation via images and films must happen for the ideas to arrive at the stakeholders.

5.6   Chatbots: Testing New Grounds with a Pinch of Pixie Dust?

David Popineau, Disney

As soon as Mark Zuckerberg announced the launch of chatbots on the Facebook Messenger platform during the summer 2016, creative agencies and brands rushed to launch their first instances on this new platform. These first chatbot experiences, mainly in the US, provided a dull conversation between motivated brands and bored consumers; only focusing on promoting products without fun nor true interaction. At Disney, I felt we should go further than this and test this new technology to see if it could drive brand likeability and serve as a truly rich experience to our audiences. Could it be a new way to advertise?

5.6.1 Rogue One: A Star Wars Story—Creating an Immersive Experience

Our first experience in chatbots was to create an advergame around the movie "Rogue One: A Star Wars Story". This chabot plays as a game and takes the users directly in the heart of the story of this first Star Wars standalone movie. For context, the story of the movie is taking place between episode III and IV of the saga, where the heroes will have to steal the plans of the Death Star, aka the ultimate arm of destruction created by the Empire. This chatbot is an immersive experience from the 1st second. The users find out their rebel competencies by answering a few questions and then go into mission to free captured-fellow rebels.

A scoring system injects tension into the whole experience. If users are captured, they will be able to ask help from their friends; which creates reasons for other users to join the experience and play the game.

This Star Wars Experience couldn't be complete without easter eggs: these little surprises or reactions from the chatbot when users are typing certain keywords or sentences from the saga. This is another way to drive interest and likeability towards the experience but was also a massive seduction tool for our Star Wars core fans.

The response from the audience was great in terms of engagement and time spent. With an average of 11 minutes in time spent, this was the most engaging item of our entire paid media plan.

5.6.2 Xmas Shopping: Providing Service and Comfort to Shoppers with Disney Fun

For our second chatbot experience, we wanted to work on Christmas as it is the biggest retail beat of the year. Looking for the best gift for parents, kids or friends can be a lot of fun but it is also very often challenge. With our Christmas Genie, we wanted to make the quest for the perfect gifts easy for our consumers. And as I was saying earlier, the shopping chatbots I had seen were no fun at all. We wanted something special, something that only Disney can do. Through a decision tree of questions, the chatbot identifies the best products for the personality of each family member. It's a personalised journey and it is a new way of shopping, a sort of digital personal shopper that takes your hand and makes your life easier. We paid a lot of attention to the immersiveness of the experience so that the chatbot reflects our Disney values of innovation, creativity and storytelling.

And the "only Disney can do" piece was in the tone of the conversation, the fun in the replies and reactions from the chatbot when the user was replying. We tested the experience in a focus group and the feedbacks were really positive. The main issue was that we struggled to get massive traffic to the experience.

5.6.3 Do You See Us?

These two experiences were great test and learn experiences. We were among the first to launch and as it is often the case, you tend to feel a little lonely in the playground. You don't get many benchmarks to adjust your proposition and Facebook is helping out but is also learning as they walk with you.

Engagement through time spent was really excellent for the two chatbots we developed. The focus group reinforced the fact that users were pleased with the experiences, and the innovation vibes they were getting from the two chatbots were really reflecting on our brands, which was one of our KPI.

The challenge was to get people to the chatbots within messenger. The users we got were mainly early adopters and during our focus group, we realised that parents audiences for instance were not aware of this new technology and got kind of scared to have 'someone' they don't know, talk to them on Messenger. The experience to get them to the bot was also challenging as they were basically served a sponsored post on their Facebook feed, which if clicked would then open their Messenger app. A real feeling that someone is taking over your device when you are not digital savvy.

The Click to Messenger format on Facebook was also just launching and was difficult to optimise as it generated high CPAs. All of these challenges we faced along the way were not surprising and were even part of the fun. But it prevented to drive scale in terms of usage.

5.6.4 Customer Services, Faster Ways to Answer Consumers' Request

Of course customer services are a much easier experience to create within chatbots. In this case, users are contacting the brand and therefore the whole burden to drive users within Messenger goes away. But on the other side, you need to align your whole customer service structure so that the staff doesn't only answer questions through emails or comments on your Facebook page, but also answers live questions from customers in Messenger. Live customer support can easily become a massive job and that's when AI must be used. We have to be careful when using AI in direct contact with customers. We don't want this intelligence to go off rails when talking to a client. But AI can be used as a filter to qualify the customer request, potentially answer a question that is very often asked, and then direct specific questions to the 'human' customer service in a seemless way.

5.6.5   A Promising Future

Above the classic customer service within chatbots, which is a massive advancement, there is no doubt that chatbots have a bright future ahead. Especially with Facebook's will to develop their millennial/Gen Z platforms such as Whatsapp, Instagram and Messenger. The key is to work on the visibility they can provide to chatbots within their platforms and we all know the massive traffic they can drive to a destination when they decide to.

Voice being the current buzz, we can also imagine these chatbots to be deported within a Facebook voice assistant? Or in any case work through your current voice assistant once Facebook Messenger is compatible with Siri and Google Assistant. In any case, chatbots must be watched as they should have some interesting development in the near future.

5.6.6   Three Takeaways to Work on When Creating Your Chatbot

Thanks to our first experiences and also as we create our upcoming projects on Facebook Messenger, we have defined 3 takeaways to pay attention to, to be successful when creating chatbots.

1. Remember What People Are Doing in Messenger

They are talking with their friends. They use GIF, Memes, they have group conversations, play games, etc. Therefore, don't try to fit an idea within Messenger. Based on what users do, what seemless and native experience can you provide within Messenger? Remember people are used to talking to humans on the platform. Therefore make the conversation as natural as possible. Don't forget to use GIF and Memes to illustrate some parts of the chatbot conversation. They are usually a hit when used properly, so have fun with it!

2. Create an Immersive Experience

Defining the tone of voice is important. Your chatbot must have a personality and a 'world' in which it lives. For instance, our Star Wars Chatbot was actually a droïd from the rebellion, therefore it was quite bossy and a little stressing. If you manage to create an atmosphere that reflects your brand or at least is very special to the experience, you will generate likeability from the user.

3. Plan the Unexpected

How do you react to an unexpected question? This is where you can make a difference. Of course, you will spend a lot of time working on your scripts and what is the general path of conversation for your chatbot. But you should almost equally spend time on unexpected things that the users will ask the chatbot. You don't want it to be stuck at the first off-script question. You should create answers that will give the illusion that the chatbot understands the question but puts the user back on the conversation rails. This is especially needed for silly questions or words. You should also think about what is core to your brand; and what users might joke about. For our Star Wars chatbot for instance, we could expect users to type "I'm your Father" to which we had prepare a funny answer and GIF. We even went to the extend of planning an emergency support if someone is seriously asking for help within the chatbot. This way, we are instantly notified by our moderation agency and can respond to the user right away in an adequate manner. The best experience will come from these unexpected little things that will make the user smile or realise that you've planned it all.

Creating chatbots is not an exact science. Experience comes as you learn to walk. Therefore I strongly advise you to not spend years planning what you chatbot will do. Instead, jump in, start small, test and learn! It will not be perfect right away but at least you'll have a first instance out there and will learn from it. Good luck!

5.7   Alexa Becomes Relaxa at an Insurance Company

Showcase: Alexa Becomes Relaxa Overview of the Development of the Skill "Smart Relax"
Bruno Kollhorst, Techniker Krankenkasse

5.7.1 Introduction: The Health Care Market—The Next Victim of Disruption?

The automotive industry is experiencing it right now, the hotel sector is in the thick of it, the taxi business and retail trade in any case. Disruptive business models are conquering one sector after the others and the pressure on former types of organisations and business models by mega platforms such as

Google, Amazon & Co, newcomers such as Dyson and start-ups like Airbnb used to be are continuously increasing. It is rather naïve to think the health care market has been spared this. In this country, it takes a little longer to establish a new health insurance company due to the particular nature of the market and the regulation by legislature, yet the planned foundation of a health insurance company by Amazon, the emergence of new players such as Ottonova and the success of platforms such as Clark allow us to guess where the journey is going for health insurance companies. The time has thus come to confront digitalisation and define it for one's self, for one's own market. This, of course, relates to internal processes, products and services in equal measure and also means nearing new technologies and channels testing them and observing how customers and potential customers deal with them. These technologies include the virtual assistant within the AI applications. This article is meant to deal with how the Techniker Krankenkasse was the first health insurance company to develop a customer-focussed service and thus approached the subject of AI. With that, to be illustrated are:

1. The considerations that led to the Alexa skill "Smart Relax"
2. The effects of this development on customers and one's own company

It is meant to be about how the acceptance and effect of AI systems can be successfully tested with simple means and in the area of conflict between data protection, customer-focussed product development and "first mover" notion without a strategic big-picture approach.

5.7.2 The New Way of Digital Communication: Speaking

First of all, a definition:

> Digital virtual assistants are a part of the AI-driven digital development of recent years. They are a software agent that, with the help of speech recognition and analysis, enables the collection of information or the completion of simple tasks and which outputs the result in a synthesis of natural-language answers. Well-known representatives of this type of software are Siri (Apple), Cortana (Microsoft), Bixby (Samsung), Google Assistant (Google) and Alexa (Amazon).

One of the most important barriers in the use of AI, the Internet of Things and smart home used to be the communication between man and machine, which was dependent in the past on interfaces such as a keyboard mouse or other manual input devices. The trend towards speech input has, however, been accompanying the triumphal procession of smartphones since 2013. With Siri, Apple was the first to implement this intuitive means of operation and thus ensured that the most important competitors caught up. Trust is established via speech and the natural means of communication helps to overcome inhibitions towards AI. The increasing possibility to apply speech control in various kinds of hardware and to thus take the leap away from smartphones should be able to prepare additional ground for the triumphal procession of the digital virtual assistant. The reach in Germany is indeed still in the starting blocks but, according to a study by Tractica from 2015 (*Source* Statista), 1.8 billion users worldwide are said to rely on this means of communication by 2021. In Germany itself, during a survey (*Source* Splendid Research, Digital Virtual Assistants, 2017) Splendid Research was able to ascertain that more than a third of the Germans already use virtual assistants, more than a third of them have a smart speaker such as Amazon's Echo (Fig. 5.12). Amazon, with their wide range of hardware, strong marketing communication and openness towards developers and the use of

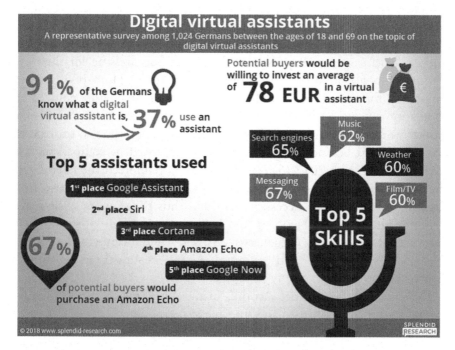

Fig. 5.12   Digital virtual assistants in Germany, Splendid Research, 2017

APIs, especially contributes towards Alexa, so that in this country, more and more users are snatching at the chance. This development should also have a corresponding positive effect on smart home and further means of application in the future. Amazon is working on integrating Alexa from ovens to cars and on becoming the user's central speech interface.

If we also take into consideration the development of Podcasts, audio streaming services such as Spotify, Deezer & Co, or online radios such as Laut.fm, it becomes clear that voice marketing will take on a greater role in the marketing of tomorrow. Voice marketing opens up entirely new possibilities for content, service and also advertising. Yet it also demands entirely new skills of editors, planners and social medians. Working with speech or transferring a communication into a natural dialogue with a virtual assistant is surprisingly not as easy as designing a visual offer.

5.7.3 Choice of the Channel for a First Case

The developments on the health care market described above and the trend towards digital virtual assistants also led to the Techniker Krankenkasse considering using AI in this way and to gather first experiences. In addition was the possibility of being the first mover in our sector to leave behind a footprint on this new channel. When researching which virtual assistant we should use, time naturally also played a role. In the middle of 2017, Amazon began to massively increase the advertising pressure for their own platform; Google Home was still in its infancy, Apple's Homepod had been announced for one year later and the remaining systems hardly enjoyed any popularity: According to a study, 67% of all potential buyers of a smart speaker would go for Amazon's Echo (Fig. 5.12). In addition to this was the fact that the AI behind Alexa was relatively advanced in comparison with others. Surveys from 2017 (*Source* Statista) then reinforced the later choice, after all, communicating with the digital virtual assistant does have to be pleasant to establish the necessary trust and thus trigger reuse. The survey by Statista and Norstat (Fig. 5.13) did indeed result in high satisfaction with the voices of all surveyed virtual assistants, yet, when it came to the issues of "pleasant", "sympathetic" and "calming", Alexa was sometimes far ahead of their competitors.

The choice of the channel was thus made. Alexa was chosen as the playground for the first test in AI and speech assistance.

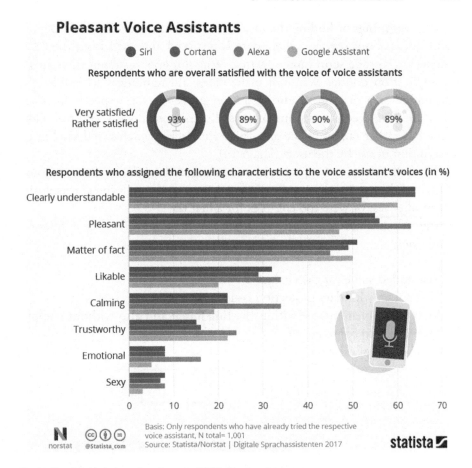

Pleasant Voice Assistants

● Siri   ● Cortana   ● Alexa   ● Google Assistant

Respondents who are overall satisfied with the voice of voice assistants

Very satisfied/
Rather satisfied   93%   89%   90%   89%

Respondents who assigned the following characteristics to the voice assistant's voices (in %)

Clearly understandable
Pleasant
Matter of fact
Likable
Calming
Trustworthy
Emotional
Sexy

0   10   20   30   40   50   60   70

N
norstat   ©①⊜
@Statista_com
Basis: Only respondents who have already tried the respective
voice assistant, N total= 1,001
Source: Statista/Norstat | Digitale Sprachassistenten 2017

statista

Fig. 5.13   Digital virtual assistants 2017, Statista/Norstat

5.7.4   The Development of the Skill "TK Smart Relax"

The next question that needed to be clarified was the question of which type of service, tool, briefing or similar do we offer? In order to determine this and to then advance the skill, a cross-departmental team was formed and a timeframe of eight weeks was defined. The roadmap was also quickly clear:

Find idea → Feasibility test → Decision for a route
   → Conceptual design → Implementation → Testing → Release

Knowing is not enough; we must apply. Willing is not enough; we must do.
—Johann Wolfgang von Goethe

At the beginnings of finding the idea, the team soon ended up in a "big picture", ideas about services where data is retrieved and used from the CRM system in real time soon came together. Why not enquire about the status of an application online? Why not make dates and reminders accessible from the online area or trigger administration processes via a speech interface? After all, a bulk of the current Alexa skills does not display any real added value but rather ranges in the area "nice gimmick", which is reflected in the description of use by the users (Fig. 5.14).

With all innovative capacity in the triangle of available time, customer focus and data protection, most of these ideas fell through the cracks during a feasibility test or were too complex for a first try.

The Alexa Skill was thus meant to fulfil the following framework conditions:

1. Real added value for the customer
2. No contact with data protection issues
3. Must be implementable within the timeframe and also without complex connection to company IT
4. Possible use of existing content

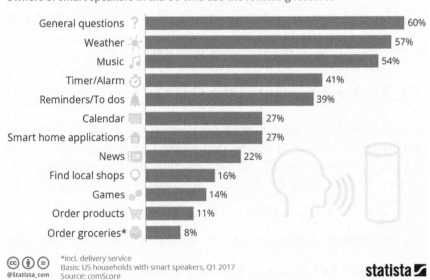

Fig. 5.14 Use of functions by owners of smart speakers in the USA, Statista/Comscore, 2017

From this consideration, the content and the subjects of our digital contents were scanned for audio usability. And out team was successful: Progressive muscle relaxation and breathing exercises for preventing stress had already been identified as content available in audio. Yet, simply playing these out over the new channel would be too inept. So let us take a closer look at the subjects of "relaxation" and "stress prevention". The studies "TK-Schlafstudie" Die Techniker, 2017, "Entspann Dich, Deutschland", Die Techniker, 2016 and the Gesundheitsreport 2017, also Die Techniker served as the basis for this consideration (Fig. 5.15).

People in Germany are under pressure: According to the TK-Stressstudie 2016, more than 60 percentage state that are frequently or sometimes

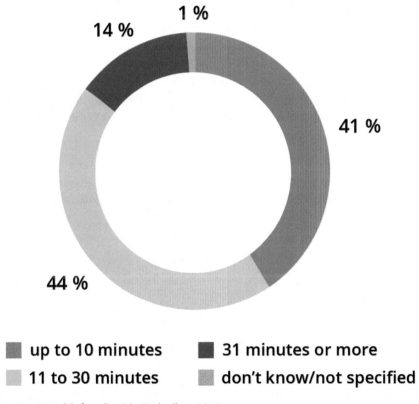

Every seventh person counts sheep for longer than half an hour

How long does it take Germany to fall asleep?

1 %
14 %
41 %
44 %

■ up to 10 minutes   ■ 31 minutes or more
■ 11 to 30 minutes   ■ don't know/not specified

Fig. 5.15 TK-Schlafstudie, Die Techniker, 2017

under stress. The health report indicates an increase in days employees are absent from work resulting from mental health issues. More than 60% of those under stress state that they feel burned out and exhausted or suffer from sleep disorders. Apropos sleep disorders: The outcome of the TK-Schlafstudie 2017: 14% of those interviewed needed 30 minutes or longer to fall asleep. It thus stood to reason to develop a solution for the subjects of relaxation, falling asleep and stress prevention.

> For the development of the skill, we used the skill builder and AWS Lambda for the backend. The method proved to be successful for us
> —Markus Kappel, developer at techdev.

Together with the communication agency elbkind and the developer techdev Solutions, who already had experience in developing skills for Alexa, the concept was elaborated and the development was started. From a customer perspective, the following questions arose for the skill:

1. What mode am I in? (After work? Sleepy? Cool down)
2. Do I need instructions on how to relax?
3. How much time do I have.

These considerations led to a methodology, firstly to create a "communicative reception hall" that with occasions (Fig. 5.16), links matching content with user moments. The occasions enable the user to be led into the two categories "methods of relaxation" and "playlists".

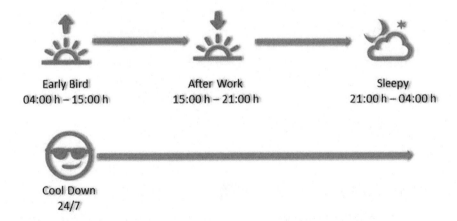

Fig. 5.16 Daytime-related occasions in the "communicative reception hall", own illustration

We first concentrate on the times of day in the morning until in the afternoon, the time after work and the phase of falling asleep. We filled the useful categories with matching content:

- **Category "methods of relaxation"**

  Meditation different methods of active meditation)
  Mindfulness (passive means of meditation to be able to perceive oneself and the surroundings)
  Progressive muscle relaxation (targeted tensioning and relaxing muscle groups)

- **Category playlists**

  Binaural beats (method of using sound to promote meditation and relaxation)
  Nature (noises from nature)
  Sleepy (music that promotes falling asleep)

Whilst the method of relaxation leaves it up to the user to choose to invest 5, 10 or 20 minutes (depending on the exercise), the playlists each contain five tracks of 10 minutes each, The sounds and songs of the playlists are coordinated such that smooth transitions are possible.

> "In the first phase of the conceptual design, we predominantly focused on the design of the conversation between skill and user. The skill offers practical help for new users to find their way around. Experienced users start their meditation exercise straight from a shortcut"
> —Bruno Kollhorst, Head of Content & HR Marketing, Die Techniker

Besides the design of the occasions, it was mainly the development of the dialogues in the implementation phase and then the testing and readjusting of them that took up a lot of time. To this end, one first needs to understand how Alex and the so-called conversational user interfaces (CUI) work (Fig. 5.17).

The following components define skills:

- **Intent:**
  Technically speaking, an intent is a function. In semantic terms, an intent is the core of the conversation, the user's intention.
- **Utterance:**

Fig. 5.17 How Alexa works, simplified, t3n

By utterances, we understand unexpected wording by the users which can serve an intent. They are thus explicitly linked up to a certain intent.

- **Slots**:
Slots are parameters with the statement of which users can specify their enquiries.

Example: "Alexa, Waste Disposal Calendar Intent) and Ask (Utterance) When (Slot) the Blue (Slot) Rubbish Bin (Slot) Will Be Collected (Utterance)".

Every skill comprises two components; on the one hand, the interaction model where intents, utterances and slots are laid down and linked up and above that, the interaction model, the actual user interface (CUI frontend). It is within the Alexa platform and helps to analyse and categorise the speech commands received. If a speech input is sent to the Alexa platform via the Echo, for example, this transforms the spoken word into text with the help of NLP. Information contained is analyses and evaluated via the interaction model.

Alon the start of the skill is very extensive in its development, but is meant to offer extensive possibilities for newcomers and pros. The most important dialogue segments to date are:

- To start: **"Alexa, open Smart Relax."**
- To jump directly into a playlist: **"Alexa, open Smart Relax and play nature."**
- To navigate within the playlists: **"Alexa, go forward."**, **"Alexa, go back"** or **"Alexa, stop."**
- To select a relaxation exercise: **"Alexa, open Smart Relax and start a relaxation exercise."**
- With limited time: **"Alexa, open Smart Relax and start a 10-minute exercise."**
- All contents can be halted with the commandos: **"Alexa, pause."** and **"Alexa, continue."**

Another train of thought was that it was necessary for the user to know the exact name of the skill and its functions to start the skill with the above-mentioned possibilities. In order to create an as positive experience as possible and to persuade the user to use the skill repeatedly, more natural ways of addressing are necessary and which also have to be tested. As the system did not recognise all speech input, the greatest compromise between natural language (pronunciation) and the set of commands that Alexa understands. The start command "Alexa, open Smart Relax" was supplemented by:

- **"Alexa, I need to relax"**
- **"Alexa, I need to recharge my batteries"**

For the hot phase of development, a constant exchange with the Alexa team of Amazon and the use of the developer portal was helpful, above all the extensive documentation, such as the "Amazon Alexa Cheat Sheet- from the Idea to the Skill" and the Speech Design Guide.

The result satisfies thousands of users to this day. Alexa became Relaxa.

5.7.5 Communication of the Skill

After the skill was developed and approved by Amazon, the work appears in the Amazon Store. To hope that it now will take off because it offers customers added value is rather short-sighted. Extensive communication is required. It makes sense, above all, to operate self-advertising on Amazon's platform as that it where the highest number of potential or already active Alexa users is. Communication is required on paid, owned and also earned channels (Fig. 5.18). To this end, a set of banners, videos and other advertising media were created and the subject of Alexa was taken into consideration in content planning. Thanks to positive reviews by the first users, the skill appears sooner or later on Amazon's radar, too and is rewarded with attention.

The inclusion in the Alexa newsletter, its being featured in the store or articles on the developer blog are only a few measures that can be valued as earned content. A suitable competition for the launch of the skill where Echos and Echo Dots could be won and which to date count among the competitions with the highest interaction rates, rounded off the communication. This way, a 360° content and advertising strategy was implemented to achieve the goals.

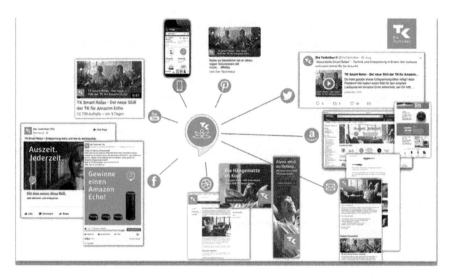

Fig. 5.18   360° Communication about Alexa skill

5.7.6 Target Achievement

The intensity of the success of the skill surprised the agency, developers, the Techniker Krankenkasse and even Amazon. In 220 days, the skill achieved more than 72,000 unique users, more than 130,000 sessions and a little more than 440,000 utterances. Cool Down, Relaxation and Playlists lie very close to each other (Fig. 5.19).

What really surprised us was the fast positive popularity on all channels. There were indeed critics who did not really greet the link between health and data grabbers such as Amazon, yet the feedback on social media channels and also in the Amazon Store was, however, predominantly positive. And in addition, the goal to be the first in the health insurance sector and to create real added value was achieved.

Besides the effects in the direction of satisfied users, a success story also developed internally. The uncomplicated and agile way of developing a new product within a short time and across departments became a frequently quoted paradigm within the organisation. The skill inspired desires in other organisational units. Sales modules with Echo as the touchable gadget and TK Smart Relax were created, lectures and health days were boosted with the skill and other ideas from other business areas are ending up with the responsible team at a high frequency. That much success demands development. The skill will experience some updates in 2018 and will end up on the Google Home Assistant with some changes to it.

Fig. 5.19 Statistics on the use of "TK Smart Relax", screenshot Amazon Developer Console

5.7.7 Factors of Success and Learnings

A significant insight from the project is the fact that no great 100% solution is needed to approach the subject of AI and virtual assistants. Significant factors that contributed to the success were:

- Top management commitment
- Greenfield approach
- Time and space for new ideas
- Mutual trust among the team, developers and agency
- Flexible, short decision paths

Some new skills, however, are required in the companies and agencies to use voice marketing correctly. The complexity of the spoken word versus the shortcomings in resent-day technology are a challenge. Making content and dialogues able to speak, making deductions for the further procedure and thus making the user experience perfect all requires a totally new way of editing. What must equally not be underestimated are data protection aspects. Especially in Germany, the scepticism towards the large Internet enterprises outshines the benefit of the offers, especially where confidential data such about one's own health is involved. The Techniker Krankenkasse was able to prove with the skill "TK Smart Relax" that meaningful added values can be created and a simple entry into the subject beyond chatbots and complex algorithms can be worthwhile on one's own CRM systems.

5.8   The Future of Media Planning

Andreas Schwabe

The international media market has been suffering for years from the self-serving and interest-driven business models of agencies. The time is ripe for a true disruption. Innovative technology companies have entered the media market with technology platforms based on algorithms. They enable transparent and efficient media planning based on AI.

1. How exactly do these new business models work?
2. What differentiates the new media mix modelling approach from the traditional agency models?
3. What are the challenges?
4. And what are the new possibilities offered for media planning—both for agencies and for advertisers?

5.8.1 Current Situation

Driven by the margin pressure in the agency scene, over the past decades media agencies have become highly creative in the advancement of existing business models. Often the budget of customers is very low. This led to agencies having to construct alternative income models. In particular, trading with media service, which implies buying and reselling of media/reach, has proven to be an extremely lucrative variant to earn additional good margins. However, this approach leads to two problems. First, the agency, which should be a neutral advisor and optimiser, leaves its advisor role and becomes a sales-oriented (re-)seller of reaches for its customers. Second, the construct leads to a lack of transparency in the media business because the margins of agencies bypass advertising customers to line the agencies' own pockets.

In 2016, the continuing discussions reached a new climax. On behalf of the Association of National Advertisers (ANA), K2 intelligence conducted an independent study about the transparency in the American media industry from October 2015 to May 2016. This study was based on 143 interviews with 150 different confidential sources, which represents a cross-section through the US media ecosystem. The results of the comprehensive 58-page study report prove that whilst all industry participants have long known it, it received too little media attention. According to the study, non-transparent business practices are part of the standard procedure of media agencies, among them hidden discounts for paid advertising volume or obscure kickback payments in form of free spots and hidden service fees. Particularly executives, who should act as role models, have been purposefully singled out to implement this approach. Even individual media buyers have been pressured in their media selection. According to the study, this affects all channels from digital to print from out-of-home to TV.

In autumn 2016, Dentsu, the fifth largest media holding worldwide, to some unintended attention in the media because it admitted to being responsible for irregularities in processing the media negotiations on behalf of its key customer Toyota.

The standard practices in the media industry and the increasing public pressure, cause a lot of discontent among advertisers. All parties involved in the market agree: they no longer want to accept the situation that rather optimising the customers' goals, agencies focus on planning and buying in accordance with their margin interests. The industry is at the point that it looks actively for solutions, for example in the form of alternative providers, who ensure a sustainable transparency and long-term planning methods solely in the interest of the customer. The time is ripe for a true disruption of the media industry.

5.8.2 Software Eats the World

Disruption describes a process, which enables a young company with less resources than the established market partner to challenge established companies successfully.

In general, established companies focus on the improvement in their products, which are already profitable, and they neglect the true needs of the market. Innovative companies use this opportunity to produce something novel and efficient, which successfully displace existing products, market or technologies and completely replace them in the end. On the lookout for alternatives for the existing business models, the large media networks, which live from outdated business models such as trading and share discounts, innovative technology companies are in the process to take this opportunity to enter the media market. This will all be made possible through the use of "intelligent systems", which solely address the customers' benefits and which create a competitive edge through the use of AI and machine learning.

"Software eats the world" is both slogan and opportunity for a sustainable revolution of the entire media industry. Through the rapidly changing technological framework conditions and digital transformation, which by now covers more and more economic sectors, this trend will develop relentlessly into data-driven decision-making processes. Data-based methods are already well-known from the performance world; however, in the media industry that is still highly offline-driven, it has not yet gotten fully established. In general, media planning focuses solely on individual channels and the focus of data collection has thus far been primarily based on online media because in online advertising analysis and algorithms have been optimised for years. A holistic attribution with planning tools based on algorithms and transparency has thus far been impossible due to the lack of technology, which also considers the data situation in offline media (TV, print, out-of-home, radio broadcast).

However, attribution without offline investments is in the end always incomplete and can easily lead to wrong decisions in budget allocation.

Visionary technology companies develop innovative products due to the continuously advancing possibilities with regard to processing power and processing speed in combination with self-learning algorithms. These products started initially as simple application at the lower end of the market and then, they rise consistently towards the top, where sooner or later they will replace the established advertising agencies completely. Elements such as automation, evaluation in real-time raise media planning and data analysis to a completely new level. New products such as media platforms are compared to traditional planning tools significantly more dynamic, more flexible and focused solely on the true needs of advertisers. This is precisely what visionary technology companies have already recognised as the true market need and which has enabled them to enter the market successfully.

Therefore, the dilemma of media agencies is that they not only have to stand up against other media agencies but also against the increasingly strong competition from cross-industry market participants. Whilst the large media holdings, which primarily receive their reason for existence by bundling the purchase power, continue to optimise their business models, new players in the market use alternative models, which are superior in optimisation and which in the long-term will completely replace the old models. Harvard Professor Clayton Christensen describes this process as "disruptive innovation" and one can say across the board that all of marketing will be facing a classic disruption. Experience shows that such developments cannot be stopped.

5.8.3 New Possibilities for Strategic Media Planning

The innovation drivers of these new players are multi-disciplinary teams of market researchers, statisticians, behavioural psychologists, mathematicians, physicists and media experts. These data science teams work continuously on algorithms, which become more and more precise, for analysis and optimisation of media investments. New, transparent business models, swifter decision-making processes and the future vision of just-in-time media make these new players attractive and effective. Blackwood Seven is one of these new players, which has already recognised those market opportunities and which has successfully monetised them. The software company developed a data-driven, automated, and self-learning platform solution that allows advertisers to plan, book and optimise media specifically focused on a KPI. Strategic media planning as software as a service (SaaS) opens up completely new opportunities for advertisers. With the help of Blackwood Seven, persons in charge of the budget can understand the effect of various media channels in interaction and they can quantify the added value of individual media investments with regard to target KPIs such as sales and new business. Because for the first time ever, the effective contributions of individual campaign elements including primarily the offline channels such as print, TV, out-of-home can be dynamically quantified with a tool, they can be evaluated objectively and therefore, they can be understood.

The innovative platform solution consists of several components, which in their interaction permit complete strategic media planning. The components cover all areas from data connection, modelling, optimisation, result simulation, visualisation, reporting, and the modulation of media.

The customer receives a complete infrastructure for its model-supported media planning. Based on this data foundation made up of internal and external variables and with the support of algorithmic modelling of all data points, various scenarios are calculated, which point out to the customer the approach for the perfect media mix. This custom-tailored modelling is dynamicised and the results are stored in the platform.

All simulations can be accessed through an individual customer interface. Therefore, the investment for the customer is low. He needs no additional hardware but merely an Internet browser. In addition, this solution is significantly faster to use than any software that runs on the customer's own computer. The customer has individual dashboards available through the web frontend. These provide a detailed insight into the latest development of the KPIs specified and into the effects of various media investments. The same web frontend illustrates the media data collected, the results of analysis and optimisation. Through the direct access to the platform, the customer can have a transparent view of its own media planning 24/7. The simulations generated with the model can be compared with customer's own results. They can be adjusted and optimised.

Persons responsible for the budget receive an unprecedented transparency and planning certainty. All parties involved are automatically interested in the actual success of the campaign and not in the maximisation of the investment.

The costs for the software can be scaled. Payment is based on defined KPIs such as turnover or sales. The customer may select between two areas, the insight part (media analytics) and the simulation part (prediction). It must be differentiated between one-time costs such as the development of a KPI model per KIP, modelling, set-up fee and onboarding along with the monthly fees to operate the platform (per applicable KPI). Blackwood Seven grows within its business model through monthly subscription fees and not through invoicing models of traditional ad agencies.

5.8.4 Media Mix Modelling Approach

The media mix modelling applied by Blackwood Seven is based on a combination of various methods. The basis is the method according to Bayes. The Bayesian statistic is characterised by consistently using probabilities or marginal distributions, which allow particularly valid results. In the course of the enormous processing capacity available today, the comprehensive data basis and use of the Monte Carlo sampling method, complex Bayesian simulations can be applied today more effectively than in the past. Bayesian modelling shows the final utility of individual media channels under various conditions, always dependent on factors such as budget, campaign period, weather, seasonal conditions and spending mood.

Media synergies (hierarchical influence various media investments have on one another) can be quantified and the effects can be maximised. Continuous model updates enable a swift response to current market developments. This is the only approach that allows transferring the complexity of the real world into the model.

5.8.5 Giant Leap in Modelling

Modelling considers various KPIs upon customer's request (Fig. 5.20). However, for modelling to be successful, it needs to be noted that the KPI is directly influenced by the media, i.e. it describes directly the behaviour of consumers.

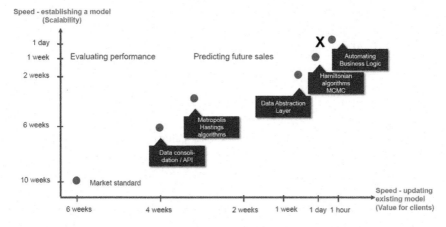

Fig. 5.20   Blackwood Seven illustration of "Giant leap in modelling"

The created model considers all media channels the customer placed. This includes unpaid media such as the customer's own homepage or his own YouTube channel. In addition, media investments of competitors and information on market changes are also considered. The customer's individual model also includes macro-economic changes, product variations, weather and other data, which describe external circumstances of the market.

A data record for all channels, covering the past three years, is needed for the initial set-up of the model. Depending on the available data basis, a daily or weekly model is created. The calculated formula value determines final utility effects, retention effects for each medium and media efficiencies in reference to the KPI. In addition, the effect offline media has on online media is considered as synergy model. Moreover, even the effect offline media and online media have on unpaid media channels can be modelled. This allows mapping indirect effects correctly.

The use of a Bayesian modelling approach offers two significant benefits: First, it is possible to integrate any possible prior knowledge from market research, a customer journey analysis or additional expert knowledge (e.g. maximum available circulation of specific media) and therefore, stake out for the model the framework conditions of the market. Second, the Bayesian approach offers a significantly more detailed result than the classic statistic. Not only can one data point (e.g. the mean value) be assigned to each parameter of the model and to each prediction but rather the entire distribution. The distribution is not only used to quantify the most probable result but also the uncertainty connected with it. This allows minimising the risks in media planning or approaching it more deliberately in order to use any potential opportunities.

As a result of modelling, the effect of the media investment and all other variables that were considered can be quantified on the modelled KPI. The return on investment (ROI) and the saturation curve of each individual media channel can be calculated based on the model. The Bayesian approach also allows pointing out the uncertainties.

The optimisation model can be used to calculate the perfect media mix for the KPI development. In addition, any existing commitments to individual media (commitments) can be considered in the optimisation of the budget distribution.

In addition, it is possible to simulate the result of any existing budget distribution and to compare various scenarios (Figs. 5.21 and 5.22).

Fig. 5.21 Blackwood Seven illustration of standard variables in the marketing mix modelling

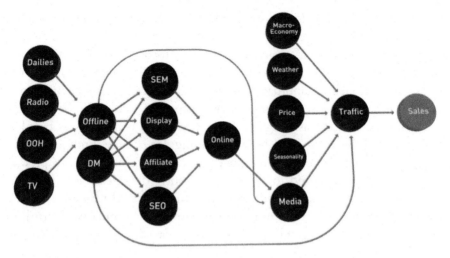

Fig. 5.22 Blackwood Seven illustration of the hierarchy of variables with cross-media connections for an online retailer

It becomes obvious: one-on-one relationships show reality only in a limited manner. Complex models, which produce multi-level effect relationships are required to map reality as precise as possible in the model.

5.8.6 Conclusion

The media mix modelling approach of Blackwood Seven is in many ways different than conventional regression models. Thus far, advertisers had no certainty in media planning and they were only able to explain the past. Today, persons responsible for budgets can simulate with planning certainty the campaign effect and understand it thanks to the machine-learning approach, which will become more precise over time. It reviews and corrects assumptions (so-called priors). The evaluation of the past and the daily model updates with the latest data allows for an exact simulation of the perfect media mix and the campaign results (always optimised to the defined KPI).

The new modelling approach allows mapping nonlinear relationships and dynamics. All important variables such as competitor information, micro-economic and macro-economic effects, customer data become part of modelling and they are considered holistically, which shows a far greater slice of reality. Cause and effect are precisely attributed whilst regression analyses requires the independence of results (maximum likelihood method), which is non-existent in time series.

So far, the mandate of a media agency was to buy as many target group contact for a fix budget as possible. The decisive parameters of the new transparent modelling approach are not the net reaches of circulation or GRPs but hard KPIs such as sales, new customers, web traffic or what else the customer specifies as his goal. The times, when planning was based on obscure Excel sheets are over. Media planning 2.0 is done through machine learning and automation, so that advertisers have a real chance to produce a holistic comparison and to show transparently the effect each individual medium produced. Model updates in real time enabled through fully digitalized processes and algorithms, which allow the formula world to learn independently, deliver results and insights of a completely new depth of detail.

These newly gained insights lead in turn to a significant gain in efficiency in media planning.

The rapid advancement of computer processor capacities and the unstoppable digitalisation going hand-in-hand with it, will bring increasingly automation and self-learning system to media planning. Media agencies must advance to bridge the gap between strategic consulting, efficient media buying, technology development and transparency for the customer. It must keep up with the speed, precision and complexity of the new systems, which implement AI in planning. Moreover, even the requirements on media planners will change. For one, they must become true data experts because data form the basis for the systems. On the other hand, media planners must be experienced media experts to develop efficient strategies and to orchestrate measures effectively.

Data procurement will become one of the greatest challenges for the media industry. Variables, processes established for years and (media) currencies will have to be re-evaluated. The uniform currency of media must be effect. The existing target conflict between advertisers, media agencies and promoters must be eliminated and an ROI consideration based on effect must be at the start of each planning process. Of course, this considers strategic brand management. And here, the human factor will continue to play a key role—at least for the next few years. We all will have to wait and see with excitement, when decisions will be made that are better than today due to the strategy with the continuously developing algorithms. It will not take too much time anymore.

5.9   Corporate Security: Social Listening, Disinformation and Fake News

Using Algorithms for Systematic Detection of Unknown Unknowns

Prof. Dr. Martin Grothe, Universität der Künste Berlin

5.9.1   Introduction: Developments in the Process of Early Recognition

The increasing digitisation of economic and public processes as well as our private lives, offers a great number of innovative and potentially beneficial features. And of course, the skilful search ("Artificial Intelligence") and the linking of relevant data through algorithms has reached further levels of information and value creation. The cyber space no longer functions merely as a parallel virtual world—it has become an inherent information and communication space.

The purpose of this article is,

- to demonstrate how, in this space, beyond IT security, other threats are increasing exponentially: Digitisation is changing fundamentally the principle of disinformation and its potential actors: a multifaceted threat for companies is emerging.
- to introduce a computational linguistics-based technology for early recognition of potential threats as a solution approach to the growing threats.

Technology-based early recognition has become increasingly important for a variety of business units and impinges on far more company divisions than corporate security. Product development, marketing and sales, communication, risk and credit management, recruiting—all can become targets of disinformation. The digitisation of communication processes offers a variety of new opportunities, but it also requires the development and sometimes overhaul of internal procedures and decision-making processes. These developments lead the way to Digital Transformation.

This article points out that relevant technologies have been tried and tested. The challenge is now to implement them and engage in their continued and sustainable development.

Digitisation is challenging entire industries. It confronts corporate functions with new and sometimes disruptive solution approaches. And the same is true for early recognition: What are the most contentious issues? Which technology will help to make a successful leap forward, towards the future?

5.9.2 The New Threat: The Use of Bots for Purposes of Disinformation

At first, a definition:

> Disinformation means the targeted and deliberate dissemination of false or misleading information. It is usually motivated by influencing public opinion or the opinion of certain groups or individuals, to pursue a specific economic or political goal.

The Internet provides all with the ability not only to become a reader and consumer of information, but also an author. The use of digital disinformation for criminal activities is tempting since online sources have become an important—if not the most important—resource for information and opinion-forming processes. Biased and deceptive information, "fake news"; have become a major challenge for politics and security and businesses.

Obviously, no one will resort to disinformation using their real name. And the digital world offers a variety of possibilities to disclose identity such as using aliases and fake identities. In cyber space, anonymity is the normality (Fig. 5.23).

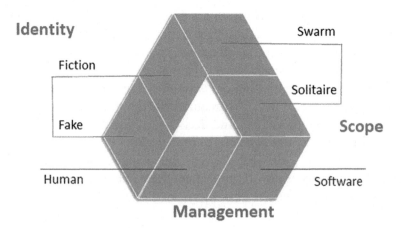

Fig. 5.23 Triangle of disinformation

5.9.2.1 Identity

Identity is one of the most important aspects in relation to the new threat. A distinction must be made between trolls and sock-puppets.

Trolls

- disrupt online communities and sow discord on the Internet
- start quarrels with other users by posting inflammatory or off-topic messages
- are isolated within the community
- try to hide their virtual identity, for example by using socket-puppets
- intent to provoke other users, often for their own amusement

Trolls are conspicuous and annoying; however, they usually do not represent a significant security threat. It's an entirely different matter with fake user accounts. Fake accounts are often called "sock-puppets".

Sock-Puppets

An additional user account to...

- protect personal privacy
- manipulate and undermine rule of a community
- discredit other users and their reasoning
- strengthen opinions and suggestions with more "votes"
- pursue general illegitimate goals

The best-known case is that of digital fictional character Robin Sage. In short, the experiment resulted in:

- Offers from headhunters
- Friend requests from MIT- and St. Paul's alumni
- More than 300 contacts among high-level military, defence, security personnel
- Classified military documents related to missions in Afghanistan
- As well as numerous dinner invitations

If your enemy knows his way around social media and social networks, information security is already at a high risk.

With digital friend requests, every hasty linking strengthens the sock-puppet's fake identity and provides her with positive network results. Simple and easy checks can reduce the risk.

Digital actors can use fictional or fake identities:

- Fake identity Design
- Identity theft

5.9.2.2 Scope

The multiplication of basic patterns results in:

- Solitaires focusing on one (or several) target (persons).
- Swarms focusing on public opinion.

Swarms can be of different size. Wealthy individuals might employ a small-scale-fan club, state institutions a "large-scale troll army". Russian activities are often described as the latter, as a state-guided digital infantry.

If the opponent controls a group of actors (sock-puppets), sentiment and opinion environments can be effectively influenced.

Businesses can also be targeted by disinformation attacks. Such an attack might:

- hurt the reputation of the company
- irritate business partners
- deter potential clients
- sidetrack suitable talents
- give an edge to competitors
- build up personal stress

All four aspects of the Corporate Balanced Scorecard can be attacked simultaneously.

5.9.2.3 Management

Targeted disinformation requires management. The increasing digitisation provides new opportunities to spread fake news, but the strategy only works for aggressors willing to engage a high number of sock-puppets.

50 years after Joseph Weizenbaum first put the software program ELIZA through the Turing Test, it has become far more difficult for humans to distinguish between human and artificial communication. The Turing test posits that algorithms should only be considered intelligent when a human interlocutor would no longer be able to determine whether he was talking to a human or to a programmed machine. Until now, this has not been achieved.

On 12 April 2016, Facebook opened the Messenger for chat bots. Human users now can ask questions for example regarding open positions or an employer directly through the messenger. AI and information retrieval are supposed to deliver the answers. Siri and Amazon Echo will follow. The Turing test has become obsolete: humans no longer see a problem in engaging in small talk with algorithms.

Bots will have significant influence on how people gather information and communicate. Bots allow for new combinations of AI and Information Retrieval/Internet Search. They can get to know their human partners in dialogue and can react conforming to profiles.

Social Bots are increasingly becoming a security risk. Non-human fake accounts are programmed to engage independently in online discussions. Via Twitter, they can also autonomously send information to manipulate and discredit other users and their opinions.

The necessary budget decreases: the new type of attack becomes available and attractive for non-state actors such as businesses and companies competing in the global market.

5.9.3 The Challenge: "Unkown Unknowns"

In addition to popular channels such das Facebook and Twitter, countless forums and blogs provide users with an enormous amount of unknown information.

In the field of Corporate Security, it is often difficult to define relevant information in advance: we are looking for something—a security risk, a threat—but we do not know precisely what we are looking for. To describe this problem, Donald Rumsfeld coined the term "unknown unknowns":

As we know, there are known knowns. There are things we know we know. We also know there known unknowns. That is to say we know there are some things we do not know. But there are also unknown unknowns, the ones we don't even know we don't know.

—Donald Rumsfeld (2002)

In a nutshell, the challenge we are confronted with is to detect weak signals long before they arise as major issues. Technological advancement offers a potential solution: using algorithms to detect issues at the earliest possible time.

Without diminishing the problems and the new threat arising due to the increased digitisation and interconnection of communication, it is worth mentioning that digitisation also offers new opportunities to confront the challenges:

- Digital noise can be used as a near real-time early warning system.
- Digital information can be used for an outside look at a company and its ecosystem including key company individuals. In taking the perspective of a malicious third party, potential weaknesses and vulnerabilities can be identified and managed.

5.9.4 The Solution Approach: GALAXY—Grasping the Power of Weak Signals

Computational linguistics and (social) network analysis are important value-adding technologies: Algorithms support content analysis by filtering through great amounts of digital content to find significant terms.

Linguistic corpora defining how often a term appears normally, exist for a variety of languages. If a term is used more often than its defined normal frequency it means that the term's significance increases. The analysis of term frequency distribution among contributions offers further guidance. In using significance and frequency analyses, computational linguistic algorithms discover relevant anomalies in a rich context without predefined search terms.

A substantive assessment of the findings demands a human touch. Nevertheless, the human mind should not be put to work on tasks that algorithms can perform: algorithms help to reduce lengthy manual approaches. They also allow for extended data coverage and real-time observations.

The described technology is superior to the popular social media dashboards that only allow to classify findings per predefined categories. The typical monitoring dashboards can count the number of absolute findings but lack any content-based indexing. which makes the method inadequate for recognising weak signals and the unknown unknowns.

Complexium's Galaxy technology offers five functions based on computational linguistic algorithms:

5.9.4.1 Discovery

Crawler and algorithms can identify anomalies in digital content. Terms are recognised and classified regarding their significance. Such an automatic exploitation of blog postings, discussion forums and other online sources allows for searches through digital content in real-time. In addition to that, the tool also enables the user to work with predefined search categories. The combination of the two approaches offers by far the best chances of discovering both known unknowns and unknown unknowns (Fig. 5.24).

5.9.4.2 Ranking

Following the classification per term significance, the tool presents a ranking overview of all terms: the daily topic ranking. The ranking shows at a glance which topics are currently found at the centre of online discussions. Additionally, the ranking can be displayed for a longer period, enabling the user to observe the development such as ups and downs of certain topics or the sudden emergence of new issues. The tool points the user towards weak signal at a very early stage. Weak signals usually appear as slow "climbers" in the topic ranking. Users can keep an eye on their development and early measures against them—if they represent potential threats—can be undertaken (Fig. 5.25).

5.9.4.3 Clustering

The topic ranking is followed by the concept-based clustering. In adapting Social Network Analysis (SNA) algorithms, the clustering reveals interconnections between groups of terms. The clustering overview shows in detail which groups of terms are more interconnected than connected with the rest of the terms. This leads to an automatic delimitation of various concept-based clusters.

Rank		Term	Change in rank
Sun Sep 10 2017	Sat Sep 09 2017		
1	1	fake	— 0
2	2	cnn	— 0
3	3	fakenews	— 0
4	5	trump	↑ 1
5	4	isis	↓ -1
6	6	rt	— 0
7	7	america	— 0
8	9	people	↑ 1
9	8	hates	↓ -1
10	10	bbc	— 0
11	12	yahoo	↑ 1
12	14	media	↑ 2
13	13	propaganda	— 0
14	16	believe	↑ 2
15	17	myanmar	↑ 2
16	20	president	↑ 4
17	21	video	↑ 4
18	25	irma	↑ 7
19	23	real	↑ 4
20	37	hurricane	↑ 17
21	24	tornado	↑ 3
22	27	stop	↑ 5

Fig. 5.24 Screenshot: GALAXY emergent terms

5.9.4.4 Mapping

In addition to the clusters, the tool generates topic maps based on a predefined list of sources to structure discussions around specific themes, companies or brands. These semantic maps show the most significant terms in relation to each other by calculating the semantic frequency of certain words. Lines of connection, font sizes and colours show at a glance term occurrence and strong coherence between given terms. The user is provided with an interactive real-time map that permits to explore the contexts of a variety of different terms (Fig. 5.26).

Fig. 5.25 Screenshot: GALAXY ranking

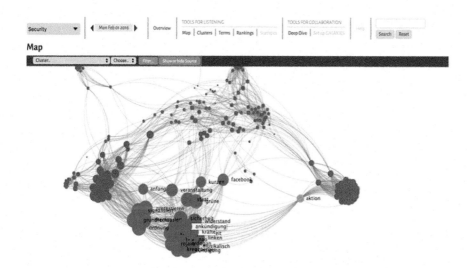

Fig. 5.26 Screenshot: GALAXY topic landscape

5.9.4.5 Analysis

As a last step, the tool's Deep Dive display helps the user to assess weak signals in terms of relevance and criticality. Provided with an overview of the sources for those significant terms shown in the ranking, clustering and mapping, the user can order and evaluate content and context of the findings. The button "assign status" enables the user to rate each finding with the possibility to earmark or forward it to other users (Fig. 5.27).

5.9.4.6 Conclusion

The increasing digitisation has generated enormous amount of data and entirely new categories which can both be of use for a variety of corporate functions. To remain up to date and competitive, businesses must engage in a wide range of transformation processes. New methods and tools to achieve this goal are already available to businesses. This article presented one such tool—the cloud-based GALAXY technology.

The GALAXY technology can support and improve processes for many corporate divisions by exploiting online content quickly and systematically.

Fig. 5.27 Screenshot: Deep dive of topics

Significant advantages are generated due to the application of innovative computational linguistic methods. This is not only interesting for Corporate Security, but also for Marketing, Communications and Employer Branding.

Thus, the GALAXY technology provides a unique possibility to recognise weak signals amid digital noise. The qualitative analysis of online sources offers an ideal starting point for more in-depth studies and a substantial analytical advantage for early detection of warning signals in a variety of company divisions, based on the following key pillars:

- Effectiveness: A detection of weak signals from relevant online sources including blogs, forums, news and review portals almost in real-time. As a "learning system" the extensive set of sources is constantly evolving.
- Efficiency: The technology makes it much less time-consuming to collect relevant information. Therefore, more time and resources can be invested in the interpretation and analysis of the results.

The GALAXY technology's explorative approach allows for a significant expansion of coverage and a systematic detection of weak signals—imperative to cope with the emerging hybrid threats.

5.10 Next Best Action—Recommender Systems Next Level

Jens Scholz/Michael Thess, prudsys AG

Recommender Systems are becoming more and more popular because they increase customer satisfaction and revenue of retailers. In general, these systems are based on the analysis of customer behaviour by means of AI. The aim is to provide customers added value by offering personalised content and services at the point of sales (PoS). In this article we first give a general definition of the task of recommender systems in retail. Next we provide an overview of the state of development and show the challenges for further research. To meet these challenges we describe an approach based on reinforcement learning (RL) and explain how it is used by the prudsys AG.

5.10.1 Real-Time Analytics in Retail

Data analysis traditionally plays a central role in retail. With the rise of the internet, smart phones, and many in-store devices like kiosk systems, coupon printers, and electronic shelf labels real-time analytics becomes increasingly important. Through real-time analytics PoS data is analysed in real time in order to immediately deduce actions which in turn are immediately analysed, etc.

Until now, for data analysis in retail different analysis methods are applied in different areas: classical scoring for mailing optimisation, cross-selling for product recommendations, regression for price and replenishment optimisation. They have been always applied separately. However, these areas are converging: e.g. a price is not optimal in itself but for the right user over the right channel at the right time, etc.

The new prospects of real-time marketing lead to a shift of the retail focus: Instead of previous *category management* now the customer is placed into the centre. Therefore the *customer lifetime value* shall be maximised over all dimensions (content, channel, price, location, etc.). This requires a consistent mathematical framework, where all above-mentioned methods are unified. Later we will present such an approach which is based on RL.

The problem is illustrated in Fig. 5.26. It exemplarily shows a customer journey between different channels in retail.

The dashed line represents the products viewed by the customer. But only those with a basket symbol attached have been ordered. In the result, the customer only ordered products for 28 dollar (Fig. 5.28).

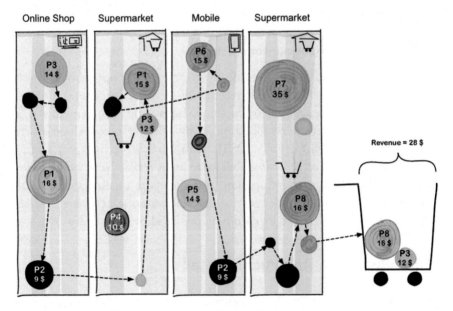

Fig. 5.28 Customer journey between different channels in retail

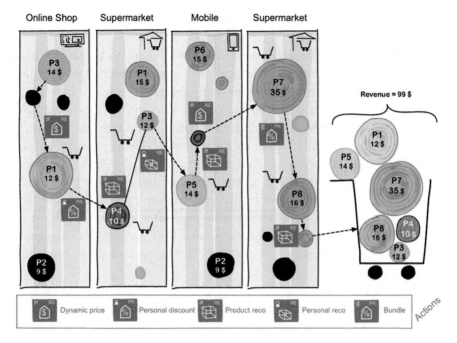

Fig. 5.29 Customer journey between different channels in retail: Maximisation of customer lifetime value by real-time analytics

Figure 5.29 illustrates for the same example the application of real-time analytics to increase the customer lifetime value (here, simply the total revenue).

Here, different personalisation methods such as dynamic prices, individual discounts, product recommendations, and bundles are used. For example, for product P1 a dynamic price reduction from 16 to 12 dollar has been applied which resulted into an order. Then a coupon for product P4 has been issued which has been redeemed into the supermarket. Then product P3 has been recommended, etc. Through this type of real-time marketing control finally the revenue has been increased to 99 dollar.

In the following we first want to examine the current status quo of recommender systems which will serve as starting point for solving the comprehensive task described before.

5.10.2   Recommender Systems

Recommender systems (*Recommendation Engines*—REs) for customised recommendations have become indispensable components of modern web shops. Based on the browsing and purchase behaviour REs offer the users additional content so as to better satisfy their demands and provide additional buying appeals.

There are different kinds of recommendations that can be placed in different areas of the web shop. "Classical" recommendations typically appear on product pages. Visiting an instance of the latter, one is offered additional products that are suited to the current one, mostly appearing below captions like "Customers who bought this item also bought" or "You might also like". Since it mainly relates to the currently viewed product, we shall refer to this kind of recommendation, made popular by Amazon, as *product recommendation*. Other types of recommendations are those that are considering the overall user's buying behaviour and are presented in a separate area as, e.g., "My Shop", or on the start page after the user has been recognised. These provide the user with general, but personalised suggestions with respect to the shop's product range. Hence, we call them *personalised recommendations*.

Further recommendations may, e.g., appear on category pages (best recommendations for the category), be displayed for search queries (search recommendations), and so on. Not only products, but also categories, banners, catalogues, authors (in book shops), etc., may be recommended. Even more: As an ultimate goal, recommendation engineering aims at a total personalisation of the online shop, which includes personalised navigation, advertisements, prices, mails, text messages, etc. Even more: As we have shown in the initial section the personalisation should be made across the whole customer journey.

For the sake of simplicity, however, we will study mere product recommendations. In what follows we consider a small example for illustration. It is shown in Figs. 5.28 and 5.30.

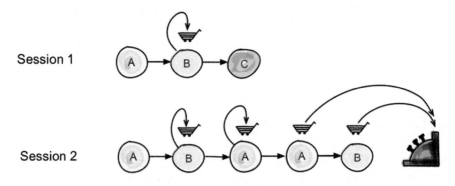

Fig. 5.30 Two exemplary sessions of a web shop

The example consists of two sessions and three products A, B, C. In the first session the products are subsequently viewed, whereat the second was put into the basket (BK). In the second session the first two steps are similar. In the third step product A was added to the basket and in the last two steps both products have been subsequently ordered. We will call each step an event. The aim is to recommend products in each event such as to maximise the total revenue.

Recommendation engineering is a vivid field of ongoing research in AI. Hundreds of researchers are tirelessly devising new theories and methods for the development of improved recommendation algorithms. Why, after all?

Of course, generating intuitively sensible recommendations is not much of a challenge. To this end, it suffices to recommend top sellers of the category of the currently viewed product. The main goal of a recommender system, however, is an increase in the revenue (or profit, sales numbers, etc.). Thus, the actual challenge consists in recommending products that the user actually visits and buys, whilst, at the same time, preventing down-selling-effects, so that the recommendations not simply stimulate buying substitute products, and, therefore, in the worst case, even lower the shops revenue.

This brief outline already gives a glimpse at the complexity of the task. It is even worse: many web shops, especially those of mail order companies (let alone book shops), by now have hundreds of thousands, even millions of different products on offer. From this giant amount, we then need to pick the right ones to recommend! Furthermore, through frequent special offers, changes of the assortment, as well as—especially in the area of fashion—prices are becoming more and more frequent. This gives rise to the situation that good recommendations become outdated soon after they have been learned. A good recommendation engine should hence be in a position to learn in a highly dynamical fashion. We have thus reached the main topic of the book—adaptive behaviour (Fig. 5.31).

We abstain from providing a comprehensive exposition of the various approaches to and types of methods for recommendation engines here and refer to the corresponding literature, e.g. (Bhasker and Srikumar 2010; Jannach et al. 2014; Ricci et al. 2011). Instead, we shall focus on the crucial weakness of almost all hitherto existing approaches, namely the lack of a control-theoretic foundation, and devise a way to surmount it.

Recommendation engines are often still wrongly seen as belonging to the area of classical data mining. In particular, lacking recommendation engines of their own, many data mining providers suggest the use of basket analysis or clustering techniques to generate recommendations. Recommendation engines are currently one of the most popular research fields, and the num-

Fig. 5.31 Product recommendations in the web shop of Westfalia. The use of the prudsys Real-time Decisioning Engine (prudsys 2017) significantly increases the shop revenue. Twelve percent of the revenue are attributed to recommendations

ber of new approaches is also on the rise. But even today, virtually all developers rely on the following assumption:

Approach 1

What is recommended is statistically what a user would very probably have chosen in any case, even without recommendations.

If the products (or other content) proposed to a user are those which other users with a comparable profile in a comparable state have chosen, then those are the best recommendations. Or in other words:

This reduces the subject of recommendations to a statistical analysis and modelling of user behaviour. We know from classic cross-selling techniques that this approach works well in practice. Yet it merits a more critical examination. In reality, a pure analysis of user behaviour does not cover all angles:

1. **The effect of the recommendations is not taken into account**: If the user would probably go to a new product anyway, why should it be recommended at all? Wouldn't it make more sense to recommend products whose recommendation is most likely to change user behaviour?
2. **Recommendations are self-reinforcing**: If only the previously "best" recommendations are ever displayed, they can become self-reinforcing, even if better alternatives may now exist. Shouldn't new recommendations be tried out as well?
3. **User behaviour changes**: Even if previous user behaviour has been perfectly modelled, the question remains as to what will happen if user behaviour suddenly changes. This is by no means unusual. In web shops data often changes on a daily basis: product assortments are changed, heavily discounted special offers are introduced, etc. Would it not be better if the recommendation engine were to learn continually and adapt flexibly to the new user behaviour?

There are other issues, too. The above approach does not take the sequence of all of the subsequent steps into account:

4. **Optimisation across all subsequent steps**: Rather than only offering the user what the recommendation engine considers to be the most profitable product in the next step, would it not be better to choose recommendations with a view to optimising sales across the most probable sequence of all subsequent transactions? In other words, even to recommend a less profitable product in some cases, if that is the starting point for more profitable subsequent products? To take the long rather than the short-term view?

These points all lead us to the following conclusion, which we mentioned right at the start: whilst the conventional approach (Approach 1) is based solely on the analysis of historical data, good recommendation engines should model the interplay of analysis and action:

Approach 2

Recommendations should be based on the interplay of analysis and action.

In the next chapter we will look at one such approach of control theory—RL. First though we should return to the question of why the first approach still dominates current research.

Part of the problem is the limited number of test options and data sets. Adopting the second approach requires the algorithms to be integrated into real-time applications. This is because the effectiveness of recommendation algorithms cannot be fully analysed on the basis of historical data, because the effect of the recommendations is largely unknown. In addition, even in public data sets the recommendations that were actually made are not recorded (assuming recommendations were made at all). And even if recommendations had been recorded, they would mostly be the same for existing products because the recommendations would have been generated manually or using algorithms based on the first approach!

So we can see that on practical grounds alone, the development of viable recommendation algorithms is very difficult for most researchers. However, the number of publications in the professional literature treating recommendations as a control problem and adopting the second approach has been on the increase for some time (Shani et al. 2005; Liebman et al. 2015; Paprotny and Thess 2016). Next we will give a short introduction to RL.

5.10.3 Reinforcement Learning

RL is an area of machine learning, concerned with how software agents ought to take actions in an environment so as to maximise some notion of cumulative reward. RL is used among other things to control autonomous systems such as robots and also for self-learning games like backgammon or chess. RL is rooted in control theory, especially in dynamic programming. The definitive book of RL is (Sutton und Barto 1998).

Although many advances in RL have been made over the years until recently the number of its practical applications was limited. The main reason is the enormous complexity of its mathematical methods. Nevertheless it is winning recognition. A well-known example is the RL-based program *AlphaGo* from Google (Silver and Huang 2016), which recently has beaten the world champion in Go.

The central term of RL is—as always in AI—the agent. The agent interacts with its environment. The interaction between agent and environment in RL is depicted in Fig. 5.32.

The agent passes into *a new state s*, for which it receives *a reward r* from the environment, whereupon it decides on *a new action a* from the admis-

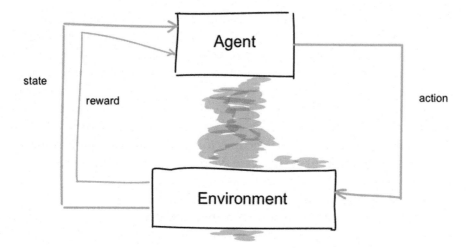

Fig. 5.32   The interaction between agent and environment in RL

sible action set A(s), by which in most cases it learns, and the environment responds in turn to this action, etc. In such cases we differentiate between episodic tasks, which come to an end (as in a game), and continuing tasks without any end state (such as a service robot which moves around indefinitely).

The goal of the agent consists in selecting the actions in each state so as to maximise the sum of all rewards over the entire episode—the *expected return*. The selection of the actions by the agent is referred to as its *policy π*, and that policy which results in maximising the sum of all rewards is referred to as the *optimal policy*.

In order to keep the complexity of determining a good (most nearly optimal) policy within bounds, in most cases it is assumed that the RL problem satisfies what is called the *Markov property*.

> **Markov property**
>
> In every state the selection of the best action depends only on this current state, and not on transactions preceding it.

A good example of a problem which satisfies the Markov property is the game of chess. In order to make the best move in any position, from a mathematical point of view it is totally irrelevant how the position on the board was reached (though when playing the game in practice it is generally helpful).

On the other hand it is important to think through all possible subsequent transactions for every move (which of course in practice can be performed only to a certain depth of analysis) in order to find the optimal move.

Put simply: we have to work out the future from where we are, irrespective of how we got here. This allows us to reduce drastically the complexity of the calculations. At the same time, we must of course check each model to determine whether the Markov property is adequately satisfied. Where this is not the case, a possible remedy is to record a certain limited number of preceding transactions (generalised Markov property) and to extend the definition of the states in a general sense.

Provided the Markov property is now satisfied (*Markov Decision Process—* MDP) the policy π depends solely on the current state, i.e. a $= \pi(s)$. For implementing the policy we need a state-value function f(s) which assigns the expected return to each state s. In case the transition probabilities are not explicitly known, we further need the action-value function f(s, a) which assigns the expected return to each pair of a state s and admissible action a from A(s). In order to determine the optimal policy RL provides different methods, both offline and online. Here the solution of the *Bellman equation* plays a central rule which is a discretised differential equation.

Once the action-value function is known the core of the policy $\pi(s)$ consists in selecting the action which maximizes f(s, a). For a small number of actions this is trivial; for a large action space, however, this may result in a difficult task. To avoid sticking in local minima it is useful not always to select actions which maximise f(s, a) ("exploit mode") but also to test new ones ("explore mode"). Here the exploration can simply be done by random selection or, more advanced, by systematically filling data gaps. The last approach is called "active learning" in machine learning or "design of experiments" in statistics.

We now turn to the application of RL for recommendations. Intuition tells us that the states are associated with the events, the actions with recommendations, and the rewards with revenues. It turns out that RL in principle solves all of the problems stated in the previous section:

1. The effect of the recommendations is not taken into account: the effect of recommendations (i.e. actions) is incorporated through f(s, a).
2. Recommendations are self-reinforcing: Is prevented by the exploration mode.
3. User behaviour changes: The central RL methods work online, thus the recommendations always adapt to changing user behaviour.
4. Optimisation across all subsequent steps: Results from the definition of expected return.

Nevertheless, the application of RL to recommendations is not simple. We will describe this in the next section.

5.10.4 Reinforcement Learning for Recommendations

The ultimate task of application of RL to retail can be formulated as follows. In each state (event) of customer interaction (e.g. product page view in web shop, point in time of call centre conversation) to offer the right actions (products, prices, etc.) in order to maximise the reward (revenue, profit, etc.) over the whole episode (session, customer history, etc.). The episode terminates in the absorbing state (leaving the super market or web shop, termination of phone call, termination of customer relationship, etc.).

To this end, we consider the general approach in RL. Basically two central tasks need to be solved (which are closely related):

1. Calculation and update of action-value function f(s, a).
2. Efficient calculation of policy $\pi(s)$.

We start with the first task. To this end we need to define a suitable state space. The next step is to determine an approximation architecture for the action-value function and to construct a method to calculate the function incrementally. For retail this is a quite complex task since we often have hundreds of thousands of products, millions of users, many different prices, etc. In addition, many products do not possess a significant transaction history ("long tail") and most users are anonymous. This leads to extremely sparse data matrices and the RL methods work unstable.

The prudsys AG is a pioneer in application of RL to retail (Paprotny and Thess 2016). For example, the prudsys Real-time Decisioning Engine already uses RL (for product recommendations) for over ten years. In order to solve the comprehensive RL problem properly and to fulfil the Markov property, over several years the prudsys AG together with its daughter Signal Cruncher GmbH have developed the New Recommendation Framework (NRF) (Paprotny 2014). The NRF follows the philosophy of RL pioneer Dmitri Bertsekas: To model the entire problem as complete as possible and then simplify it on a computational level.

Here each state is modelled as sequence of the previous events. (i.e., each state virtually contains its preceding states.) For our example of Fig. 5.32 the three subsequent states of Session 1 are depicted in Fig. 5.33.

In the example the first event of Session 1 is a click on product A. Thus, it represents state s1. Next, the user has clicked on product B and has added

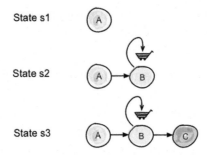

Fig. 5.33 Three subsequent states of Session 1 by NRF definition

it to the basket. Thereby, the sequence **A click→ B in BK** is considered as state s2. Finally, the user has clicked on product C. Hence the sequence **A click→ B in BK→ C** click forms the state s3.

By this construction, the Markov property is automatically satisfied. We now define a metric between two states. It is based on distances between single events from which distances between sequences of events can be calculated. This metric is complex by nature and motivated by text mining. For this space we now introduce an approximation architecture. Examples are generalised k-means or discrete Laplace operators. In the resulting approximation space we now calculate the action-value function incrementally. Within the NRF actions are defined as tuples of products and prices. This way products along with suitable prices can be recommended.

The correctness of the learning method is verified by simulations. For this purpose, we learn in batch online mode over historical transaction data and in each step the remaining revenue is predicted and compared with the actual value. The results of simulations show that the NRF ansatz is suitable for most practical problems.

Next we consider the second task: The efficient calculation of policy $\pi(s)$, i.e. the determination of the maximum value of f(s, a). We therefore need to evaluate the action-value function f(s, a) for all admissible actions a of state s. Moreover, often the choice of actions is limited by constraints (e.g. suitable product groups for recommendations and price boundaries for price optimisation). These constraints are often quite complex in practical applications.

To overcome these problems, in very much the same way as for the state space, for the action space a metric was introduced. Based on this metric, generalised derivatives have been defined which allows to calculate the optimal actions analytically and efficient. At the same time, through a predicate logic a syntax for generic definitions of constraints has been developed. A predicate processor transforms the constraints in a unified internal form which then is used in policy evaluation. Nevertheless, complex constraints limit the action space drastically and may lead to long calculation times. The acceleration of this process is an interesting task for further research.

In the result, the NRF enables the efficient implementation of combined product and price recommendations. Additionally, an extension by further dimensions like channel and time is intended. In this way, the vision of Section 1 could soon become real.

5.10.5 Summary

Recommender systems go far beyond the scope of product recommendations only: they can increase the customer value over the whole customer journey. This requires a new mathematical thinking. Instead of just analysing historical behaviour of customers their interplay with recommendations should be modelled. A proper tool for this purpose is RL. In this article we have discussed the application of RL to recommender systems by presenting a powerful new approach. This leads to interesting mathematical problems which should encourage further research in this area.

5.11 How Artificial Intelligence and Chatbots Impact the Music Industry and Change Consumer Interaction with Artists and Music Labels

Peter Gentsch

5.11.1 The Music Industry

Music by its nature has always been a non-tangible product. However, the medium on which this product was distributed and used by consumers has changed substantially over the last few decades and centuries. Simultaneously to the growing physical industry in the 1920s, the first US radio station opened in 1921. To that time, big major labels and disc manufacturers ignored early signs of success by those emerging stations which lead to a big decline in their market share and finally ended in most of the labels and disc manufacturers being bought by rising radio stations. Radio stations mainly podcasted live concerts and events, whereas gramophone discs were only used for occasional listens at home. This led to a historical decline of turnover for the recording industry of 94.3% from 1921 to 1933. Later, during the so-called Rock'n'Roll-Revolution the Federal Communication Commission (FCC) decided, that the restriction of radio licenses in every state of the USA is cancelled. In the following, many independent radio stations got the right to podcast music, for which they mainly used music recorded on vinyl discs. This replenished the record industry and led to a new boom of newly emerged music styles, especially Rock'n'Roll which gave its name to the movement (Gentsch et al. 2018). This boom lasted until the late 1970s with major labels such as CBS Columbia, Warner Music, MCA and EMI owning almost the whole value chain of the music industry, including music agencies and instrument construction companies. The introduction of the Compact Disc (CD) in 1982 by electronic giants Sony and Philips as well as emerging music television shows led the industry to new highs. During this time, many big companies from different industries invested heavily into the music industry which led to many company fusions and a new line-up of the three major record labels Sony Music Entertainment, Universal Music Group and Warner Music Group that are still in the industry today. With the turn of the millennium and the rise of filesharing platforms such as Napster, the global turnover again decreased dramatically (Gentsch et al. 2018).

Streaming is a technology to receive a continuous stream of data over the Internet that enables the recipient to directly access the transmitted data without having to wait for the download of full files. Commonly, it is used to access audio or video data.

In the following, the three biggest music streaming services Spotify, Apple Music and Amazon Music are described and compared. Especially Spotify is analysed in detail regarding its business model and technical background. It is explained how Spotify utilises Peer-to-Peer systems to provide efficient music streaming and how AI impacts the generation of automated song and artist recommendations.

5.11.1.1 The Technology Behind Music Streaming

Caching is an often-used process of Internet platforms to provide stutter free services. In the case of Spotify, often played songs of a user are downloaded to the user's cache storage, thus the songs are not needed to be re-downloaded when the user listens to them the next time. To decrease buffer times and unwanted stops during streaming, Spotify uses a combination of client servers in a P2P (Peer-to-Peer) network to exonerate its own servers and to provide large scale and low latency music-on-demand streaming. In a P2P network, every user serves as a node in the system and processing work is partially directed to each user's computer to improve the overall processing power and to distribute work in the most efficient way. For the actual data transmission, Spotify uses the Transmission Control Protocol, short TCP, which is a network communication protocol designed to send packets of data via the Internet. Firstly, TCP is a very reliable transport protocol because data packages that got lost on the way to the receiver can be re-requested. This avoids missing data which can result in audio and video glitches, i.e. stuttering playback. Secondly, TCP is friendly to other applications in the same network that also use TCP, therefore, multiple applications running simultaneously do not hinder each other's data transfer. Thirdly, Spotify's P2P network and TCP benefit each other as the streamed files are shared in the network and are therefore easy to re-access.

Spotify's decision from which source the song will be streamed, i.e. the server, the local cache or the P2P network depends on the amount of data the client has already at disposal and whether the song selection is a random hit or a predictable track selection. If the song is a frequently played song, the data will be drawn from the local cache. If the song is not stored at the local cache, the client reaches out for the Spotify server and the P2P network. This ensures that the needed data packets can be accessed in time. However, some Experts state that the remaining 39% of plays are random hits that cannot be predicted. This occurs when a user clicks on a random song which is not in the predicted, and therefore prefetched, order of songs. In case of a random hit, the TCP is used to quickly load approximately 15 seconds of the requested song from the Spotify server. Simultaneously, the player reaches out to the P2P network to access peers who have parts of the song stored in their cache and draws the data packets from them. In case that none of the peers has any data packets of the song the client stops uploading data to the P2P network for other clients in order to use more bandwidth to load the song from Spotify's own server.

Streaming services rely on recommendations to provide their customers with new content. To enhance the user experience, the recommended content should align with the user's personal taste. Spotify's recommendation model is a machine learning hybrid approach to generate automated recommendations. It uses Collaborative Filtering, Natural Language Processing (NLP) as well as neural networks that analyse raw audio tracks to generate personalised recommendations that are meant to meet each user's specific taste.

5.11.1.2 Collaborative Filtering

Collaborative Filtering, short CF, is the most commonly used recommendation system for streaming services. In the case of Spotify, users cannot "like" or give a rating to songs, therefore, the algorithm uses other information to search for similar tastes between users, which is the stream count of songs and additional information, e.g. how songs are placed in users' playlists and how often artist pages get visited. Furthermore, Spotify creates a unique vector for every single user and song and recommends based on the similarity of these vectors.

5.11.1.3 Profiling Through NLP

In addition to CF, Spotify also uses NLP to profile music. Spotify scans the Internet for blogs, news and articles to learn how they describe and define certain artists. This information is integrated into the taste profile of each user and helps to identify other artists and songs that are similar to the ones the user likes. This usage of NLP is based on written text; however, NLP is not limited to that. Especially personal digital assistants that work via voice control take advantage of NLP to process spoken words into information.

5.11.2 Conversational Marketing and Commerce

5.11.2.1 Conversational Marketing

First of all, it is important to define the term conversation. Linguistically, a conversation is driven by cooperation which includes a direction, a meaning and clear goals of every participant. Molly Galetto, Vice President of Marketing and Communication at Belgium-based NG Data, describes Conversational Marketing as a feedback-oriented approach to marketing, which is used by companies to drive engagement, develop customer loyalty, grow the customer base, and, ultimately, grow revenue. The difference to traditional content marketing is its direction. Rather than talking at the customer, companies talk with the customer, i.e. it is an interactive exchange, a two-dimensional conversation. This two-dimensional conversation is vital for companies as they get access to valuable customer data that they did not have before. Good communication and service in an in-person interaction increases customer loyalty and leads to higher revenue. The same applies for Conversational Marketing, with the only difference that it is a virtual conversation. The customer's interests are identified within the conversation and used to generate personalised information matching the customer's needs. If the provided information resonates with the respective customer, they are more likely to con-vert and generate future business for the company. Furthermore, mobile applications, chatbots and voice assistants allow a continuous customer service support, 24 hours a day, 365 days a year. This becomes more and more important because customer service has become part of the marketing as an integral element of conversations. In general, Conversational Marketing, compared to traditional content marketing, follows a long-term strategy that is personalised for each customer.

5.11.2.2 Chatbot Platforms

Chatbots, or virtual assistants, are defined as computer programs, that converse with users in natural language. Their field of application is very broad and ranges from entertainment purposes to education, business, the query of information and commercial purposes.

As the use of machines becomes an increasingly important part of people's lives and as the number of machines grows every year, people desire to communicate with them in a way that is more similar to the communication they use towards other people, i.e. natural language. Chatbots are a tool to meet this desire and make Human-Computer Interaction (HCI), the interaction between human users and machines, more natural and human-like. HCI advantage over common human-computer interaction is the real-time responsiveness of the program (Unbehauen 2009).

5.11.2.3 Standalone Solutions

Other than chatbot platforms, where chatbots from many different companies are implemented into a host-app, standalone solutions are chatbots or other conversational tools that are em-bedded into a corporate website or app. There are only a few big chatbot platforms, however, every company can build its own standalone solution specifically tailored to its needs. Although the portfolio of a standalone bot is limited to a company's products and services, the tailored User Interface (UI) often works better than the standardised UI of chatbot platforms. IBM also offers a bot-building service called Conversation that runs on IBM Wat-son Application Programming Interface (APIs) and that does not require any programming know-how from users to create chatbots for various applications such as customer engagement, education, health or financial service.

5.11.2.4 Voice Recognition Software

The Cluetrain Manifesto states that markets are conversations and their members communicate in natural and open language which cannot be faked. Almost twenty years later, voice recognition software goes beyond written text-based NLP as digital voice assistants can communicate by using human voice. In contrast to standalone chat bots, with many companies utilising the technology, only a few big players such as Amazon, Microsoft, Apple and Google have developed digital voice assistants for commercial purposes. The human voice has become a new interface to operate machines as it is described in Voicebox Technologies's 2006 patent application called "System and method for a cooperative conversational voice interface". To utilise this technology, voice recognition software has to be embedded into matching hard-ware, featuring a microphone and a speaker such as smartphones or smart speakers. Up to date, the biggest and most popular smart speaker is Amazon Echo which accounted for 70.6% of the total use of digital voice assistants in the USA in 2017. Google Home, the second most popular one, being far behind with 20.8%. Hereafter, Amazon Echo and especially its voice assistant Alexa is described further.

5.11.3 Data Protection in the Music Industry

AI and chatbots, of companies such as Spotify, gather data in order to create user profiles, recommendations and other services. Furthermore, companies use data processing to optimise their overall business and profitability. However, the collection of data as well as its processing is usually regulated by law, but different countries approach this issue differently which means that some companies, that are incorporated internationally, might not be obliged to respect the data protection laws of the countries where the customer is located.

The European Union currently applies two laws that specifically address data protection. Firstly, the Data Protection Directive 1995/46/EC (Section 2.1) which is the basis of EU data protection, and secondly the e-Privacy Directive 2002/58/EC (Section 2.2) that was specifically designed to address the protection of personal data in telecommunications. On May 25th, 2018, the new General Data Protection Regulation (GDPR) will officially come into force and replace the Data Protection Directive 95/46/EC. It was designed to further harmonise data protection across Europe and also focuses on data privacy regarding border-crossing inter-national organisations. The key change is the extended jurisdiction of the GDPR, which specifies that it applies to all processors and controllers that process the personal data of data subjects living in the EU regardless whether a company is located in the EU or not. Companies outside of the EU that process personal data of EU citizens must install an official representative in the EU. Moreover, the regulation awards more rights to the data subject such as the "Right to be For-gotten" that entitles them to let their data be erased by the controller, or the right to access their data in which case the controller must provide a copy of the personal data free of charge. For consumers, this is a big improvement regarding data transparency and data protection.

In the USA, there is no common general law that applies for every federal state. Instead, each federal state elaborates own laws that sometimes overlap with other federal states, however, sometimes also vary substantively or even contradict each other. One of the most important regulations is the Federal Trade Commission Act (FTC Act), a federal law aiming to prohibit de-ceptive and unfair actions both online and offline to protect consumers' personal data and their online privacy. In conclusion to this patchwork approach of regulations, each federal state may handle data protection differently as compared to the EU's holistic approach for all member states.

At the 19th National Congress of the Communist Party of the People's Republic of China in Beijing, in October 2017, the government decided that the development of technology and innovation will be one of the country's four growth drivers for the next ten years. According to Sarita Nayyar et al., Chief Operating Officer at the World Eco-nomic Forum LLC, China will transform into a consumer-driven development model with less than five companies controlling all consumer data by 2027. In June 2017, the new

Cyber Security Law came into force that is substantially similar to the GDPR regarding the rights of data subjects, however, it leaves room for the interpretation of certain terms. For instance, the law states that network operators are not allowed to disclosure, alter or destroy personal data without the consent of the person the data is gathered from and that such information is forbidden to provide to third parties. Yet, the very definition of a network as defined by Article 76 of the Cyber Security Law, is a system of computers and other relevant devices that are capable to collect, store, transmit, exchange and process information, which also applies to many private computer networks. Moreover, all critical data that is generated in China have to be stored in China. Furthermore, what critical data and those operating it, called Critical Information Infrastructure Operators (CIIOs), exactly is, has yet to be defined in context of the new law. These circumstances make it increasingly complex for international businesses to operate in China.

Overall, the global data protection landscape is very complex and requires deep understanding to fully grasp. Fortunately, the EU puts data security to the centre of attention with its GDPR that promises more security and transparency for consumers and less bureaucratical complexity for companies. In addition, China's new Cyber Security Law shows the increasing importance of data governance in the rising consumer market that is China, whilst the USA still lacks a compre-hensive and holistic regulation for all its member states. How these new regulations will affect the global music industry in detail remains to be seen.

5.11.3.1 Qualitative Expert Interview

The purpose of the interview was to gain deeper insights of the music industry's changes due to AI and chatbots from an expert's point of view. In contrast to the quantitative survey, which was designed to get consumption insights from a consumer's point of view, the qualitative expert interview focused on industry-specific topics. The interview was of semi-structured nature and the questions were formulated in an open-ended way to allow input for further relevant and specific knowledge. Initially, there were eight separate questions which, during the interview, thematically overlapped and therefore are not listed separately in the following report.

Cherie Hu is an entrepreneurial journalist who focuses on innovative technology in the music industry and is based in New York. She holds a diploma in Piano Performance from the Juilliard School and graduated from Harvard University with a bachelor's degree in Statistics. Further-more, she works as tech columnist for Billboard and is also music columnist for Forbes. Additionally, Hu contributes to the Harvard Political Review, Music Alley, Cuepoint, Inside Arts and more. She has a deep understanding on how AI, Chatbots and other innovative technologies transform and shape the music industry. Hu received the Reeperbahn Festival's inaugural reward for Music Business Journalist of the Year 2017.

5.11.3.2 Music Discovery Through Streaming Services

The way in which AI changes the consumption of music today is most noticeable regarding Streaming services, e.g. Spotify. The use of algorithmically generated song and artist recommendations has become a habit and the algorithm even accounts for outliers in taste, that non-algorithmic recommendations simply ignored. Hu exemplified this point by describing a consumer, who likes both Lady Gaga and James Brown. These two artists are very different in their sound and usually appeal to two different target groups as they are from different generations. Services like Spotify can account for this diversity in taste by creating individual and extremely granular user profiles and thus generate much more diverse recommendations.

5.11.3.3 Music Consumption Becomes More Reactive

As many users are engaging with Spotify's Discover Weekly, the algorithm processes this user data to generate even more playlists. By doing so, users are constantly fed with new music and artist recommendations without the need to actively search for new content. Users simply select whether they like those automated recommendations, which again is a data input for their unique user profile.

According to Hu, Matthew Ogle, Product Manager at Instagram and former Product Director at Spotify, stated at a presentation during the Sónar Music Festival in Barcelona, 2016, that over 8000 artists receive more than 50% of their streams through Discover Weekly. A few thousand artists achieve even more than 75% of the streams with Discover Weekly. She further states that this development is very beneficial for smaller artists that otherwise would be unheard due to a lack of exposure. From a consumer side, the range of artists that users listen to increases every year since Spotify started to publish these figures in 2013/2014. Due to the volume of content that is being fed to the audience the users spend less time on average with a single artist, thus, with the many algorithmically generated recommendations, their listening behaviour is more di-verse. Hu estimates that listening on Spotify has become approximately 40% more diverse over the recent years and underlines that diversity is now Spotify's main product. Consequently, and although smaller artists might benefit from unusual high exposure, it is becoming harder for artists to develop a loyal fanbase on the platform.

5.11.3.4 Limitations and Challenges of AI and Chatbots in Music Streaming

According to Hu, the big question that streaming services and other tech companies try to figure out, is how to contextualise their services. For instance, if a user likes to listen to up-tempo Electronic Dance Music (EDM) whilst exercising, the recommended music should be generated according to the situation the user is in, in this example EDM for exercising. As of now, 65 contextualisation is in the beginning phase of development and common services are not yet capable to contextualise. Spotify's recent approach to address this matter is its Mood Playlists, which play music in the mood of the user. However, these playlists have to be selected manually and are curated by humans, not algorithms. Moreover, Discover Weekly, which is algorithmically generated, cannot pick up human feelings and memories that make a user play a specific song because it is important for them on a personal level. Spotify can measure the action, the selection of the song, but is not yet capable of emotional understanding, i.e. the reason that leads to the action. This seems to be the current limit of Spotify's recommendation algorithm, which makes it not the all-end answer to music discovery. Inherently built into this, is an assumption about how people discover music and what they are looking for. Present AI has not quite mastered the context awareness yet.

5.11.3.5 Transforming Role of Music Labels

One has to distinguish services such as Distrokid from real music labels, be it a major label or just a smaller indie label. Online distribution services such as Distrokid only provide distribution to online platforms and shops but do not offer any marketing activities, whereas music labels offer many different services including distribution, marketing, promotion and public relations. Hu states, that as long as artists want to focus solely on the art and do not want to get involved with the business side of the industry, there will be a place for music labels. However, traditional music labels need to adapt to recent changes and adjust in their business model. She further criticises the lack of data-driven decision-making in the music industry. Yet, coming up with unique ways of how to utilise the data has become crucial for music labels as all of the major labels receive the same data from streaming services. In the future, it will be part of music labels' strategic advantage to handle the data to establish a market advantage towards competitors.

Furthermore, labels shift to a customisable service model regarding the work with artists. According to Hu, in the past, traditional contracts with artists often were 360° deals that covered everything from music production, marketing and sales as touring. To compete with new emerging online distributors, labels more openly offer individual services which makes the contracts way more flexible from a legal point of view. By this, music labels try to stay competitive whilst artists become more flexible in how they can sign a label/service deal.

5.11.3.6 Chatbots Have Yet to Mature

Although Hu expresses that chatbots hold big value and potential, she still thinks they need to mature even further to become applicable in the music industry. The idea of direct communication with fans is not new to artists, in fact, artists have collected the phone numbers of very loyal fans to send them news and updates for years. Hu states, that even today, this is still a very effective form of communication. As of now, chatbots are not commonly used as the automatically generated text messages of chatbots are not sufficient in the way most artists and managers want them to be formulated.

5.11.3.7 The Voice Becomes the New Interface

The human voice changes the way people consume music as it becomes the new interface to control devices and services via voice command. According to Hu, this new form of control interface directly impacts music labels because they must find a way to place their artists so that they are the first thing a consumer thinks of if they want to listen to music. Then, consumers would tell the digital assistant to play music from this artist. Opposed to ordinary streaming this requires more action from the consumer's side but makes the process more human-like, which ultimately adds value to digital assistants.

The idea of AI with human voices is applicable to many fields of usage, e.g. journalism. An uprising technology in human voice simulation is Lyrebird. ai that is currently being developed at the University of Montreal. Lyrebird. ai enables users to upload voice recordings which is analysed by an AI. After processing, the user can arbitrarily write a text into a box that the AI will output in the voice of the uploaded sample audio. Regarding journalism this is highly controversial because it poses questions of the validation process.

5.11.3.8 Globalisation of the Music Industry and Collaboration with Other Industries

The Internet and streaming services, especially Spotify, largely account for the increasing globalisation of the music industry. One of the reasons for this globalisation is the interconnectivity between artists who introduce other artists to a wider audience by adding them to their personal and public playlists. Despite the increasing interconnectivity, Spotify has problems to enter new markets because in some countries there are already other established streaming services. Hu also sees the future of the music industry in the collaboration with other industries such as fashion or video games that implement music into their products or services. By working closely together with those industries, the music industry tries to maintain its current growth-phase. Especially the video game industry shows big potential, that is not yet capitalised.

5.11.4 Outlook into the Future

After 15 years of economic decline, the music industry has finally seen its first year of growth in 2015. The industry structure is constantly shifting and the way in which people consume music has changed several times over the last few decades. Both, the quantitative survey and the qualitative expert interview affirm the initially stated hypothesis that AI-driven applications lead to increasing interaction between consumers, artists, music labels and streaming providers. In addition, the thesis confirms the assumption that streaming services are of tremendous importance to the current music industry, notably Spotify, as it has disrupted the industry substantially. AI has revolutionised the industry and transformed it into a digital and globally interconnected business.

This transformation is not limited solely to the music industry, as AI pushes Conversational Marketing and Commerce to the centre of attention in the online retail market. NLP, smart recommendation systems and personalised customer service via chatbots will continue to develop and the expected future growth of smart devices hosting digital assistants confirms this trend further. In general, marketing and commerce will shift from a one-directional to an omni-directional information flow which will produce even more data. To utilise this ever-growing pool of data, it will require more data scientists and more efficiently working algorithms.

Notes

1. http://www.businessinsider.de/statistics-on-companies-that-use-ai-bots-in-private-and-direct-messaging-2016-5, accessed on 29 Sept 2016.
2. http://www.spiegel.de/netzwelt/web/microsoft-twitter-bot-tay-vom-hipster-maedchen-zum-hitlerbot-a-1084038.html, last accessed on 26 Sept 2016.

References

Accenture. (2016). *Customer Service Transformation Innovative Customer Contact and Service*. Munich.

Andreessen, M. (2011). Why Software Is Eating The World. *The Wall Street Journal*. http://www.wsj.com/articles/SB10001424053111903480904576512250915629460. Published on 20 Aug 2011.

Annenko, O. (2016). *Wie Grossunternehmen von Chatbots profitieren können*. Online. http://www.silicon.de/41626347/wie-grossunternehmen-von-den-chatbots-profitieren-koennen/.

Arbibe, A. (2017). The Challenge of Data Protection in the Era of Bots. Retrieved May 28, 2017, from https://blog.recast.ai/data-protection/.

Aspect. (2017). Customer Service Chatbots and Natural Language. Retrieved April 29, 2017, from https://www.aspect.com/globalassets/microsite/nlu-lab/images/Customer-Service-Chatbots-and-Natural-Language-WP.pdf.

Beaver, L. (2016). The Chatbot Explainer: How Chatbots are changing the App Paradigm and Creating a new Mobile Monetization Opportunity. In Business Insider Intelligence. Retrieved from http://www.businessinsider.de/what-are-chatbots-a-new-app-and-mobile-monetization-opportunity-2016-9?r=US&IR=T.

Beuth, P. (2016). *Twitter-Nutzer machen Chatbot zur Rassistin*. Online. http://www.zeit.de/digital/internet/2016-03/microsoft-tay-chatbot-twitter-rassistisch.

Bhasker, B., & Srikumar, K. (2010). *Recommender Systems in E-commerce*. Noida: Tata McGraw-Hill.

Braff, A., & Passmore, W. J. (2003). Going the Distance with Telecom Customers. *The McKinsey Quarterly, 4*, 83–93.

Brewster, S. (2016). *Do Your Banking with a Chatbot*. Online. https://www.technologyreview.com/s/601418/do-your-banking-with-a-chatbot/.

Christensen, C. (2016). *Disruptive Innovation*. Online. www.claytonchristensen.com, http://www.claytonchristensen.com/key-concepts/. Published on: Not specified.

Christensen, C., Raynor, M., & McDonald R. (2015). What Is Disruptive Innovation? *Harvard Business Review, 93*(December), 44–53.

Der Kontakter, Der Deutsche Mediamarkt krankt, in: Kontakter 31/2015, Published on: 30/07/2015, p. 16 (2015).

Dole, A., Sansare, H., Harekar, R., & Athalye S. (2015). Intelligent Chat Bot for Banking Systems. *International Journal of Emerging Trends & Technology in Computer Science, 4*(5), 49–51.

Egle, U., Keimer, I., & Hafner, N. (2014). KPIs zur Steuerung von Customer Contact Center. In K. Müller & W. Schultze (Eds.), *Produktivität von Dienstleistungen* (pp. 505–545). Heidelberg: Springer Verlag.

Elder, R., & Gallagher, K. (2017). What Social Media Platform do consumers Trust the Most? The Digital Trust Report – Business Insider Intelligence. Retrieved May 30, 2017, from http://www.businessinsider.com/the-digital-trust-report-insight-into-user-confidence-in-top-social-platforms-2017-5.

Elsner, D. (2016). *Chatbots mit Banking-Potential.* Online. http://www.capital.de/meinungen/chatbots-verfuegen-ueber-banking-potenzial.html.

Gentsch et al. (2018). How Artificial Intelligence and Chatbots Impact the Music Industry. *Research Paper, 3*(1). HTW: Aalen Germany.

Gronau, N., Fohrholz, C., & Weber, N. (2013). Abschlussbericht "Wettbewerbsfaktor Analytics-Reifegrad ermitteln, Wirtschaftlichkeitspotenziale entdecken" Ergebnisse einer explorativen Studie zur Nutzung von Business Analytics in Unternehmen der DACH-Region, Potsdam 2014.

Günther, V. (2016). *Dentsu Japan gibt Unregelmäßigkeiten bei Toyotas Mediageldern zu.* www.horizont.net, http://www.horizont.net/agenturen/nachrichten/Media-Tansparenz-Dentsu-Japan-gibt-Unregelmaessigkeiten-bei-Toyotas-Mediageldern-zu-142966. Published on 22 Sept 2016.

Hafner, N. (2016). Sprachidentifikation und Sprachanalyse auf dem Vormarsch. *Contact Management Magazine, 4,* 24–25.

Hill, J., Ford, W. R., & Farreras, I. G. (2015). Real Conversations with Artificial Intelligence: A Comparison Between Human-Human Online Conversations and Human-Chatbot Conversations. *Elsevier, 49,* 245–250.

Hoong, V. et al. (2013). The Digital Transformation of Customer Services. Whitepaper. Deloitte Consulting. Online. https://www2.deloitte.com/content/dam/Deloitte/nl/Documents/consumer-business/deloitte-nl-the-digital-transformation-of-customer-services.pdf.

Iyer, B., Burgert, A., & Kane, G. C. (2016). Do You Have a Conversational Interface? *MIT Sloan Management Review.* Online. http://sloanreview.mit.edu/article/do-you-have-a-conversational-interface/.

Jannach, D., Zanker, M., Felfernig, A., & Friedrich, G. (2014). *Recommender Systems: An Introduction.* Cambridge University Press, 2010.

K2 Intelligence, An Independent Study of Media Transparency in the U.S. Advertising Industry. https://www.ana.net/fileoffer/index/id/industry-initiative-media-transparency-report-offer. Published on June 2016.

Liebman, E., Saar-Tsechansky, M., & Stone, P. (2015). DJ-MC: A Reinforcement-Learning Agent for Music Playlist Recommendation. In *Proceedings of the 2015 International Conference on Autonomous Agents and Multiagent Systems* (pp. 591–599). Istanbul.

Mathur, A. (2017). Program Your Chatbot to Handle "Long-tail" Questions With Watson Conversation and Watson Discovery. IBM – The DeveloperWorks Blog. Retrieved July 18, 2017, from https://developer.ibm.com/dwblog/2017/chatbot-long-tail-questions-watsonconversation-discovery/.

Paprotny, A. (2014). *A Novel Optimal Control Framework for Recommendation Engines with Data-Driven Approximation Architectures.* Chemnitz: prudsys AG.

Paprotny, A., & Thess, M. (2016). *Self-Learning Techniques for Recommendation Engines.* Basel: Birkhäuser.

Price, B., & Jaffe, D. (2008). The Best Service Is No Service: How to Liberate Your Customers from Customer Service, Keep Them Happy, and Control Costs. San Francisco: Wiley.

prudsys AG. (2017). *Unsere Lösung – die prudsys RDE.* https://prudsys.de/loesung/. Accessed 25 Feb 2017.

Reichheld, F. (2006). The Ultimate Question: Driving Good Profits and True Growth. Boston: Harvard Business School Press.

Ricci, F., Rokach, L., Shapira, B., & Kantor, P. B. (2011). *Recommender Systems Handbook.* Heidelberg: Springer.

Sauter M. (2016). *Trend "Conversational Commerce": Bots ersetzen Apps.* Online. http://www.futurecom.ch/trend-conversational-commerce-bots-ersetzen-apps/.

Schnitzler, C. C. (2013). Vom Call Center zum Customer Care Center – Fit für die Echtzeitbetreuung des Online-Kunden. *Marketing Review St. Gallen, 3,* 64–73.

Service Excellence Cockpit. (2017). https://service-excellence-cockpit.ch/en/home-2.

Shani, G., Heckerman, D., & Brafman, R.I. (2005). An MDP-Based Recommender System. *Journal of Machine Learning Research, 6,* 1265–1295.

Silver, D., & Huang, A. u. a. (2016). Mastering the game of Go with Deep Neural Networks and Tree Search. *Nature, 529,* 484–489.

Simmet, H. (2016). *Individualisierter Service durch Chatbots: Die neue Welt der digitalen Kunden-Kommunikation.* Online. https://hsimmet.com/2016/06/02/individualisierter-service-durch-chatbots-die-neue-welt-der-digitalen-kunden-kommunikation/.

Sokolow, A. (2016). *Sind Chatbots das nächste grosse Ding?* Online. http://mobil.n-tv.de/technik.Sind-Chatbots-das-naechste-grosse-Ding-article17437.

Steiner, A. (2016). *Künstliche Intelligenz, Die Bot-Revolution geht los.* Online. http://www.faz.net/aktuell/wirtschaft/netzwirtschaft/unternehmen-setzen-auf-chatbots-chancen-risiken-14175914-p2.html#lesermeinungen. Accessed 7 June 2016.

Sutton, R. S., & Barto, A. G. (1998). *Reinforcement Learning: An Introduction.* Cambridge and London: MIT Press.

Weidauer, A. (2017). *Do-It-Yourself NLP for Bot Developers.* Online. https://conversations.golastmile.com/do-it-yourself-nlp-for-bot-developers-2e2da2817f3d#.ys5nj1rc8.

Part V

Conclusion and Outlook: Algorithmic Business—Quo Vadis?

6

Conclusion and Outlook: Algorithmic Business—Quo Vadis?

6.1 Super Intelligence: Computers Are Taking Over—Realistic Scenario or Science Fiction?

6.1.1 Will Systems Someday Reach or Even Surmount the Level of Human Intelligence?

We all know Hollywood's horror scenario from the film Matrix: A super intelligent computer system enslaves us humans and simulates our reality: The matrix.

Maybe we also shared the excitement with Will Smith and the humanoid robot Sonny in "I, Robot" on their mission to save the world. Yet, how realistic are such scenarios?

Everyone is talking about artificial intelligence (AI). Is it possible that we are on the brink of the breathrough of a machine super intelligence that is superior to us by miles? And how dangerous would that really be for us?

The fact is: A film about a supercomputer that supports us in our daily and professional life will not provide enough drama and action for a holywood story it would seem. We should not allow ourselves to be influenced by fiction, one which pay with a fear that exists for just as long as the fascination for an intelligence we have created.

And yet one question occupies many of us: "What will happen when we make ourselves replacable?"

© The Author(s) 2019
P. Gentsch, *AI in Marketing, Sales and Service,*
https://doi.org/10.1007/978-3-319-89957-2_6

We humans are acting in an uncontrollable environment. Through constant interaction with our environment, we are learning more and more, mostly without even noticing it.

To do this, we firstly have to be able to perceive our environment. Step by step, we are getting to know the meaning of this perception. We get to know our mother's voice when we are still in her womb, for example, yet the significance of this person only becomes clear step by step.

We therefore initially classify an object. We effortlessly test out our environment. By dropping toys, we get to know gravity. We learn that the hot food cools down all by itself if we wait long enough. This means that as early as at the age of two we have a good intuition of physical correlations in our world and how they ineract with us. We also classify increasingly more objects and assing different properties to them. This is how our common sense is developed and we are able in a certain way to predict situations such as "if I drop the glass, it will break". This ability accounts for a large part of our intelligence.

With further development, we can abstract this classification of objects. The abstraction makes it possible to compare different objects or even situations that objectively have nothing in common. By doing this, we can transfer strategies that we have successfully learned in a situation to a different situation. Our ability to transfer is a further key pillar of our intelligence.

How much sense it makes, however, to derive with our brain more about the way our brain processes from research data and precisely how this research data can be depicted at all, is another, very interesting topic of discussion.

How is our intelligence to become manifested with machines?

There is software already available that is far superior to humans in some areas. In 1996, IBM's Deep Blue defeated the reigning world champion in chess for the first time. 20 years later, in 2016, AlphaGo won at the more comlex Japanese version Go and these are only the famous examples.

The rules of the games were implemented into both systems, i.e. added into the system and trained for many years. The algorithms both systems use analyse the situation of the game and decide in favour of the strategy branch with the highest probability of success. Machines build up this strategy tree bit by bit during training. Similar to a human one would think, but simply a machine.

Yet, the great difference is that the same systems would be a complete and utter failure at "Ludo". Even the first move would be impossible, as the rules of the game would first have to be implemented by programmers. And even if both systems were taught the rules of the game, they would not be able to transfer the strategies to the new game. And it would also not be possible for them to differentiate between short-term tactics ans long-term strategies. For games like chess or Go, that does not really matter. But all the more so if we want to discharge the systems into the rough world.

Expert systems nowadays are thus already superior to humans in very naroow areas, but general intelligence with abstraction processes and transfer skills of what has already been learned, as a human-level AI system would demand, has not been achieved in the slightest.

Almost all of today's commercial successes of AI systems can be lead back to supervised learning algorithms. To this end, the systems are shown huge, already classified amounts of data. On the basis of this evidence, the system then automatically adapts the Verknüpfungsgewichte between the individual points of representation of the problem (the formal neurones). This way, individual sub-aspects of the solution are emphasised more than others. Finally, the system puts the solution together and ideally, translates the solutions from representational coding into a form that can be analysed by humans.

The comparison with sample solutions helps the system to evaluate its own result. By way of penalties or rewards, the system sees whether the learning process brings about the desired result or not. Similar to a pupil, the system is given a penalty or a reward: The principle of reinforcement learning.

The next step in emancipating the systems towards human-level AI are unsupervised learning algorithms that work in the use case. This is about unsupervised learning like with children that explore their surroudings and learn to interact with them. Here, despite current small breathroughs, research is still at square one.

As of late, there has been promising progress in the field of unsupervised learning. In 2017, the research group around Anh Nguyen from the University of Wyoming succeeded in producing synthetically generated high resolution images of volcanoes, buildings and animals. Yet, even during the training of these "Plug & Play Generative Networks", much already classified trainig data was taken. To this day, no researcher has succeeded in anything similar from mere raw data.

The problems researchers face today are as multifaceted as the field itself.

There is thus no known representation known to date that enables machines to sufficiently extract the results to apply what has been learned outside the training context. Until now, networks only abstract very superficially. For example, a specially trained network recognises animals in an image due to the high vegetation in the background—irrespective of whether there actually is an animal in the image or not. That logically leads to many false positive results. Concept learning, in which we humans are true masters from birth, is a huge problem for machines.

To date, there are no known efficient communication symbols for the human-computer interface. Indeed, the AI community has been abe to celebrate remarkable accomplishments of late in the field of machine speech recognition and translation, which everybody uses, for example, in YouTube substitles or with the Google Translator, yet machines do not understand the spoken word like we do. Thus the direct learning of machines for systems has been hardly possible to date. The correlation between facts, figures, targets, strategies and communication must continue to be implemented system- and problem-specifically. And the way things are looking, that will stay that way for quite a whilst yet. Even the summarising and presentation of results in formats comprehensible for humans is a great problem for many systems and has to be developed for each system individually.

Learning algorithms are extremely resource-intensive. An extreme amount of computing power and time is needed to train a system adequately, as the entire network has to be re-simulated for every symbol, quasi each new fact. And to date, there has been a lack of a machine-episodical memory or a long-term memory, meaning that the computer forgets everything it has learned hitherto when a new learning process is completed.

"Learning to learn" is certainly the decisive mantra for the next intelligence for the next level of maturity. Today, people are still trying to define the best learning algorithm for the system. In the future, AI systems will find the best way to learn for themselves. On the basis of a kind of meta learning process, we delegate as it were the determination of the ideal learning algorithm. This kind of AI autonomous learning goes far beyond the learning paradigms of today's machine learning. The "general problem solver" could in this way also universally beat the world chamion in chess, Jeopardy, Go and "Ludo" by always learning for itself the best solution algorithm.

Another problem is reasoning in line with common sense. A computer only knows facts that are explicitly specified and accessible. For us humans, implicit knowledge is a matter of course. When we compose a legal text, we know that colloquial expressions are out of place in it. This knowledge and the framework conditions resulting from it for the further processing of the information has had to be explicitly and problem-specifically implemented in the machines up till now.

AI is also a firm part of current research in robotics. Almost all problems are multimodal and cannot simly be transferred into one target function for machines. Facebook and DeepMind are indeed, working on a physics-based virtual environment to train such systems. But there is no system to date that is comprehensive enough to implement the demands on multi-tasking that our environment makes of us.

For example, self-driving cars do not recognise people as intelligent beings with their own home range and repertoire of strategies, but as an obstacle. The interaction with the environment is inadequate to this day. The defensive driving style resulting from that is still far from the optimum of possibilities.

In summary, it can be said that this super intelligence will come due to the rapid development and technological scaling. The question as to "when" is certainly difficult to answer. Each advance uncovers new questions and obstacles. A precise answer to this question according to the current state of research is not yet possible. An incredible amount has already happened. Some things are already possible that were only conceivable in sience fiction ten years ago. But there is still an incredible lot to do. And on the way there, increasingly more progress that we can already use for ourselves will be made. There is no field where the correlation between basic research and science and industrial application is as close as in AI. If we once take a look behing the backend scenes, some of us would be amazed at how significantly our technological landscape is already affected by AI and how much of that we already use.

If we compare various studies and expert statements, the tipping point to super intelligence is taxieren at between 2040 and 2090.

It is certain that we are on the brink of groundbreaking technology that will continue to significantly influence all of our lives and already does today. In the future, we will interact with AI systems very intimately, be it in everyday life or in our professional life. As these systems are developed to improve our life circumstances and to maximise our performance, we should not give into the fear of substitution by software. Human-level AI by no means means the creation of a new intelligent machine species will successively eliminate us from many areas of life. In fact it means that we reach the next level of human performance, with AI systems as our vehicle.

This general problem seeker and solver of super intelligence would then also mean the highest level of maturity of algorithmic support for companies. The vision of the more or less deserted and self-operating company would become reality. In order to prevent a full loss of control, it would have to be ensured that humans lay down and monitor the framework and conditions of the AI-based "learning to learn" system. This also includes the control of the red OFF switch that is frequently seen as a safety anchor. Yet, a self-learning AI system will also learn to understand such switches and how to switch them off. Otherwise we will actually run the risk of being mastered by systems sooner or later—hasta la vista, baby!

6.2   AI: The Top 11 Trends of 2018 and Beyond

Besides the development towards super intelligence, there are at present a multitude of developments in the field of AI. I the following, the key trends that have the greatest impact on business are summarised compactly:

1. **AI first:** Analogue to the "mobile first" mantra, particularly with companies such as Facebook, Microsoft and Google "AI first" prevails: No development without investigating and utilising the AI potentials. At this stage, that is certainly also a sure overvaluation due to the immense hype. At present, a downright arms war is taking place among the AI applications of the GAFA world. The M&A is equally interesting in the field for AI and febrile at the same time. Similar to mobile, AI will increasingly become a matter of course in the years to come, so that the adjunct "First" will disappear. In any case, this "AI first" mantra of the digital giants, coupled with the corresponding making available of knowledge and codes, will be a push in AI for many other industries and companies.

2. **AI will not really become intelligent, yet nevertheless increasingly important for business**: The discussion about the question as to whether and when AI is really intelligent is as old as it is unsolved. The analogy of neuronal networks suggests the intelligence claim of AI on the basis of the apparent reproduction of the human brain. Yet, even massively switched neuronal networks in parallel do not represent the human brain. To this date, how the brain really works is unexplored, how creativity can actually be generated and reproduced.

   Thanks to the immense increase in computing capacities, AI systems are increasingly creating the impression of human intelligence, because they are able to interrelate and analyse huge amounts of data in not time at all and, in this way, make good decisions autonomously. A human could never interrelate these huge, heterogenous and distributed data sets. Thanks to the AI-based reasoning of these data universes, seemingly innovative and creative results can also be generated, whereby only existing information—even if immensely large and complex—can be analysed. Even the much-quoted and discussed deep learning is not really intelligent in this spirit. In the same way, the software that can develop new software itself is conditioned and determined by the original intelligence of the original developer.

   From a business perspective, the discussion about the real intelligence must, however, have an academic appearance. After all, the quasi intelligence that simulates human intelligence increasingly better helps to support important business processes and tasks or to even perform them autonomously. For this reason, the AI development of today will change business rapidly and sustainably when it comes to intelligence, despite the not really existent quantum leap.

3. **Specific AI systems**: The dream of general AI systems independent of functions and sectors has to be dreamed for another whilst. This general intelligence shall remain the grandeur of humans for now. IBM's Deep Blue was indeed able to beat the former chess world champion Kasparow mressively, but will have great difficulty in defeating the Korean world champion in the board game Go.

   In contrast, an increasing number of domain-specific AI systems are being successfully developed and established: Systems for certain functions such as lead prediction in sales, service bots in service or forecasts of validity. This narrow intelligence will increasingly support corporate functions and also replace them.

4. **AI inside—embedded AI**: AI is bing integrated in more and more devices, processes and products. This way, AI is more frequently managing the leap from the AI workbench to business. Examples are the clever Alexa by Amazon, the self-driving car, the speech-controlled Siri by Apple or the software that automatically detects, classifies and addresses leads. The label "AI inside" will thus become more and more a given. After all, almost any physical object, any device can become smart through AI.

5. **Democratisation of AI**: Despite the immense potential of AI, only a few companies use technologies and methods of AI. This is frequently associated with the lack of access to skills and technologies. Frameworks such as Wit.ai by Facebook and Slack by Howdy alleviate the simple development of AI applications by way of modules and libraries. With tools like TensorFlow (machine learning) or Bonsai (search as a service), somewhat more sophisticated AI applications can be programmed. So-called AI as a service providers go one step further. DATAlovers, for example, provides AI methods for the analysis of business data as a service. The AI services AWS (Amazon) cover cloud-native machine learning and deep learning for various use cases. Cloud platforms such as Amazon's AWS, Google's APIs or Microsoft Azure additionally enable the use of infrastructures with good performance to develop and use AI applications.

6. **Methodical trend deep learning**: Back to the roots—just more massively: Many examples (e.g. the victory over the Korean world champion in Go, sales prediction) impressively show the potential of deep learning. The interesting thing about this trend is that the methodical basis

is anything but new. Neuronal networks that have been in discussion since the 1950s represent the basis. Thanks to the new IT infrastructures with good performance, these neurona networks can now be switched in massive parallel. Whereas there used to be two to three layers of neuronal networks, today, hundreds of layers can be switched and computed. That is not a new method in principle, but the better performing and scaleable interpretation of a famous method (the Renaissance of neuronal networks). A quasi higher intelligence is developed by this massive parallelisation.

7. **More autonomy—fewer requirements**: Unsupervised and reinforcement learning on the move: Today, a good 80% of all AI applications are based on so-called supervised learning. Training data is required for learning—who are the good guys, who are the bad guys? The algorithm learns discrimintating and differentiating patterns. This approach continues to be excessively relevant as the training data available is growing immensely thanks to the Internet and big data. In the past, there used to be bottlenecks and great efforts in generating the corresponding training data. Nevertheless, the room for expectations and solutions is given to a certain extent. When it comes to acquiring patterns in "unlabelled data", e.g. acquiring automatic segments from a data set, so-called unsupervised learning is applied. Higher autonomy in terms of the given input also enables so-called reinforcement learning. With reinforcement learning, we learn from the interaction with a dynamic system without determining explicit examples for the "right action". The control of operating robots is a typical reinforcement problem. A control system is optimised such that the robot preferably no longer falls over. However, there is no teacher to say what the "right" motor control is in a situation.

   Due to the higher degree of autonomy and of innovation content of the possible results, these methods are of particular interest for business. Due to the greatly increased computer capacities and AI infrastructures, they will be increasingly applied.

8. **Conversational Commerce as a driver**: Similar to the Internet of Everything, the increasingly important Conversational Commerce will be fuelled by the dramatically increasing number of connected smart devices as well as the necessity and imagination of AI. Conversational Commerce facilitates the optimisation of customer interaction by way of intelligent automisation. The target of overriding importance is to lead the consumer directly from the conversation to purchasing a product or service. This includes, for example, the processing of payment methods, drawing on services or also the purchasing of any products.

In these cases, messaging and bot systems are increasingly applied, which have speech- and text-based interfaces that simplify the interaction between the consumer and the company (Amazon Alexa, Google Home, Microsoft Cortana, etc.) with this, the entire customer journey from the evaluation of the product over the purchase down to service can be optimised through greater efficiency and convenience. Besides algorithms that control via keywords and communication patterns, AI is increasingly applied to learn systematically from the preferences and behavioural patterns. This not only holds true for the personal assistants and butlers on the consumer side of things, but particularly for the service and collaboration bots on the company's side of things. Consumer and company bots will increase the demand for AI sustainably.

9. **AI will save us from the information overkill**: There are enough facts and figures about how rapidly the amount of information is increasing dramatically. The big data analyses in turn produce new data. The information overkill is impending. But this is exactly where AI will help by intelligently filtering, analysing, categorising and channelling. NLP (natural language processing) will become more efficient so that speech and text can be increasingly processed automatically. AI-based filter systems will progressively help to not only confine the flood of information but also automatically distil added values from the flood of information.

10. **Besides the business impact of AI, the economic and social change caused by AI is increasingly becoming the topic of conversation**: After the megatrends Internet, mobile and the IoT, big data and AI will be seen as the next major trend. The digital revolution is also being called the third industrial revolution. Similar to the industrial revolution 200 years back, the radical change triggered by digitalisation will bring about change in both technology and (almost) all areas of life. AI and automation will progessively reduce working hours and also substitute jobs. This is discussed critically in the following final Sect. 6.3.

11. **Blockchain meets AI**: The subject of blockchain is discussed vigourously in the context of the Bitcoin currency. It is, however, also of significance perspectively for AI-based marketing. Due to the monopoly-like market power, the AI landscape dominated by the GAFA world or the BAT world in China (Baidu, Alibaba, Tencent) bears the risk of lacking transparency of the used data and AI models in particular that can be misused for manipulative purposes. Do you trust all answers and recommendations by Alexa, etc.? "The bot market is estimated to grow from $3 billion to $20 billion by 2021" (https://seedtoken.io). On the one hand, the Alexa models could be acting not in yours but in Amazon's

spirit. On the other hand, the interface could also be hijacked, meaning that you also receive recommendations that do not match your structure of preferences. This is exactly where the concept of a decentral, transparent and non-manipulable blockchain mechanism could help against the key AI and big data approaches.

At the same time, it is all about the three AI levels:

- (Big) data layer
- Algorithm/AI layer
- Interface layer

With today's centralised solutions, we have to trust the integrity and safety of the data. If the data for training AI is biased or intentionally falsified, the results of the AI model are also falsified. Even if the data and algorithms are "clean", the recommendations to the AI interface can still be manipulated. The user has no transparency about what is happening behind the curtain of a centralised approach.

Users can be rewarded by cryptographic tokens that can be moneterised by providing their data on appropriate marketplaces. An example of this is the Ocean Protocol (https://oceanprotocol.com). The protocol as a decetral exchange protocol provides an incentive for the publication of data for training AI models. With products such as Nest, Fitbit or other IoT services, the data sovereignty and use lies with the respective producers. On the one hand, the user is not rewarded for providing their data; on the other hand, there is no guarantee that the providers are using the best AI models on the data. The Ocean Protocol thwarts this:

- Data integrity (transparency of the source of data)
- Clear ownership (of the respective users and "donors")
- Cost-efficient settlement for purchase/rent

An energy AI model optimised on the basis of the nest data could, for example, now be made available to other users via a marketplace, who can feed and use the model with their data. As there is also clear ownership with regard to the AI model, an adequate set-off or reward is safeguarded as per the blockchain approach.

The SEED network can be named as an example for this. SEED is an open, decentral network in which all bot interactions can be managed, examined and verified. The network also ensures that the data fed into the AI system can be allocated to a data owner, who can be recompensate for it.

If a provider not only developed an ideal AI model for hone energy consumption on the basis of the nest data, but also a (chat)bot that asks you at regular intervals: "Hey, are you feeling too hot or cold in your house at the moment?" Your replies are fed directly into the AI model—and after all, it is your data. Why should you not be reimbursed for that? After all, it makes the AI models better and adds to the data repository. SEED thus secures your proprietary rights in the blockchain. Another advantage is the greater trust in the authenticity and credibility of the (chat)bot you are interacting with.

This blockchain AI approach could represent a counterbalance to the deadly spiral of the AI of the GAFA world. The GAFA companies, on the one hand, start off with an already extremely high degree of AI maturity; on the other hand, they invest billions of dollars in the expansion of AI technology and hire the best data scientists. Furthermore, they generate more and more data via platforms that, in turn, facilitates ever better AI models. In a self-reinforcing process, the AI full stack companies (they even build for AI optimised processes) on the basis of the platform and scale effects increase their lead more and more and thus create uncatchable market entry barriers.

Over time, increasingly more data could flow into the blockchain "publicly and democratically" and thus put the market power of the GAFA world into perspective. This way, increasingly open marketplaces for data and AI models can be forecasted.

6.3 Implications for Companies and Society

The mantra "algorithmics & AI eat the world" at the beginning of the book responded to the immense disruption potential for companies and society at an early stage. The interesting question is what will be eaten, who eats and who will be eaten.

Algorithmic business is the subject-matter and result of the so-called current fourth industrial revolution. In the three industrial revolutions of the last 200 years, the economy and society emerged strengthened, despite the consistently prevailing fears: Higher productivity, more wealth, better educational background, longer life expectancy, etc. Can we now also expect this happy end with the fourth industrial revolution?

Whilst during the second industrial revolution, the likes of factory workers, who were at risk due the automation of production, saw their salvation in the driving of trucks—true to the motto "vehicles will always be driven by people"—the question is increasingly posed as to which professions will be made up for by AI-endangered workers. Will this industrial revolution also lead to more wealth and productivity like the revolutions before did? These challenges as well as questions of ethics and privacy will shape the AI discussion in the future.

Interestingly, the subject-matter of this fourth industrial revolution is not really that new—it is about digitalisation. It was all about digitalisation back in the micro-electronic revolution of the 1970s and 1980s. Due to the immense potentials for change and design for business, the current revolution is not about gradual but radical change.

Social criticism is currently being fuelled by the division of society forced by digitalisation. Digitalisation acts as a booster for winners and losers: The rich continue to win, the poor continue to lose. The danger is in the augmentation of the digital two-tier economy.

What are the economical and social consequences exactly? There is a consensus to a large extent in theory and practice that algorithmics and AI will change the working world in the long run. About a half of today's jobs will no longer exist I 2030. A topical World Economic Forum Report predicts that more than five million jobs will be lost to AI and algorithmics in the next four years. The Mckinsey Global Institute (2013) estimates that 140 million full-time jobs could be replaced by algorithms by 2025. According to calculations by McKinsey, algorithmics and AI data will automise the work performance of ten million financial experts and lawyers by 2025. What used to take experts days to do is now done by computer programs in minutes.

Figure 6.1 accordingly illustrates the clear reduction in working hours per week.

What will we do with the newly acquired free time? How can we displace value added chains in a meaningful way? How can redundant jobs and activities be transferred to and turned into new value added chains? How can we turn the time acquired through substitution into innovations and creativity and use it?

These key questions for our society are becoming a matter of considerable debate.

As Jenry Kaplan said in 2017: "AI does not put people out of business, it puts skills out of business". Employees will thus have to apply their skills elsewhere or learn new skills. Richard David Precht sees the development rather critically. He not only sees the economical with scepticism but also the psychological aspects. The phenomenon of "self-efficacy", the meaningful feeling of getting somewhere doing something because you have done it yourself, is in danger. The question is whether this self-efficacy can also be realised and lived in the newly acquired window of free time, or whether

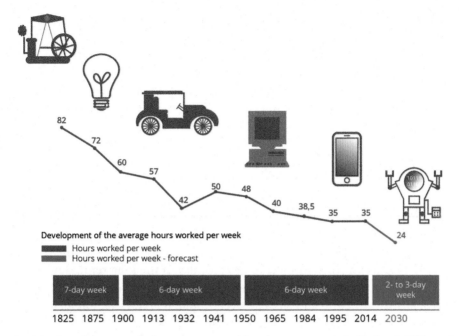

Fig. 6.1 Development of the average working hours per week (Federal Office of Statistics)

digitalisation makes the world void of meaning, work, experience and feeling.

Algorithmic business implies an intense automation of processes in and between companies. The future challenge for companies will be to find the right degree, the right balance of automation. This way, customers will accept a booking process of a flight being performed by Conversational Commerce mechanisms. No customer here will miss an empathetic conversation with the service agent or a sophisticated storytelling approach. Smart customers will increasingly use bots that control this booking process more or less autonomously themselves. But there are also customer situations in which human-to-human communication as a socialising and trust-forming element can be critical for success. A full automation of the customer journey across all touchpoints in the spirit of a bot-to-bot interaction does not appear to be constructive in the short to medium term.

For companies, algorithmic business means a change in paradigm to data-driven real-time business. The increased potentials through big data and AI are also associated with these challenges, however. If companies succeed in systematically collecting and processing the data and in implementing corresponding measures, potential benefits—as shown in the best practices

(Chapter 5)—can be achieved in the shape of optimised customer experience, reduction in costs and increase in turnover.

Despite the potential for operationalisation and optimisation o algorithmic and AI, it must not be forgotten that economic actors can still also behave emotionally and irrationally at times. Consumers and decision-makers will not allow themselves to be conditioned to become homo-economicus—i.e. rationally dealing actors in the future either.

> As we all seek automation in operations, we must not lose sight of the fact that our customers are human.[1]

The time has come to place customers at the beginning of the digital value added chain. AI makes it possible for every company to build up an automated and strongly personalised customer relation, to bind them more closely to the company and secure their loyalty in the long term. Some technologies such as social media bots are, in fact, not yet fully mature, yet, an efficient infrastructure and a data-driven implementation requirement in the company must first be developed; and that takes time.

Algorithmics and AI can play out their strengths in the automatic collection, generation and analysis of data. With clear interaction schemata and standardised communication, the communication can also be automated in the shape of drip campaigns and content creation. The creative design of communication and campaigns or the explanation of consumer needs will also still remain he domain of human intelligence for now. The extent to which these activities will be taken over by AI in the medium or long term will have to be awaited. The first promising AI applications already create pieces of music or draw artworks today, and thus demonstrate the potential for creativity of modern AI.

As the digitalisation of processes, communication and interaction will also increase in the future, the associated amount, speed and relevance of data will continue to increase. Accordingly, the approaches of algorithmic business described will play an increasingly important role in the competitiveness of companies.

The fact that this automation is not only a goal pursued by companies, but that it also corresponds with customer motivation and thus makes the breakthrough in Algorithmisierung and automation of company-customer interaction seem probable is emphasised by the Mckinsey study on this:

> By 2020, customers will manage 85 percentage of their relationship with an enterprise without interacting with a human.[2]

It is not about the mechanistic and technocratic electrification and digitalisation of processes. Algorithmics and AI have the potential to also question existing processes and business models fundamentally and to come up with completely new business processes and models. True to the motto of the former Telefónica CEO Thorsten Dirks: "If you digitalise a crappy digital process, you will have a crappy digital process".

Companies that understand an implement accordingly algorithmics and AI are the winners of tomorrow. These core competencies will decide over competitiveness and are already doing this today. Amazon, for example, is not a marketplace nor a retailer, Google (or Alphabet) is not a search engine or media outlet—first and foremost, both are algorithmic businesses, that collect, analyse an capitalise data perfectly. Companies need this skill to gain future competitive advantages themselves. Business AI enabled companies are anxious to interanlise this skill via intelligent software and services and to turn it into competitive advantages.

Frequently, technologies are overestimated in the short term and underestimated in the long term. In addition, we frequently lack the imagination as to the speed at which these developments change businesses and societies.

Famous experts have, for example, estimated that it will take at least 100 years for AI to beat the world champion in Go—reality showed it happened much faster.

Last but not least, a few false estimations of technology developments that show how frequently and blatantly potentials of technologies and innovations have been falsely estimated.

The fact that the technological developments (big data, AI, IoT, Conversational Commerce, etc.) described in this books are developing exponentially and not linear and that we, as entrepreneurs and society are still standing at the bottom of the exponential ascent, makes it clear that the actual potential still lies ahead of us The algorithmic business has only just begun and has immense potential that none of us can reliably forecast at the end of the day.

Those who can imagine anything, can create the impossible (Alan Turing 1948).

Notes

1. Simon Hathaway, Cheil Worldwide 2016, https://www.retail-week.com/analysis/…and…/7004782.article, last accessed 10 July 2017.
2. Baumgartner, Hatami, Valdivieso, and Mckinsey 2016, https://www.gartner.com/imagesrv/summits/docs/na/customer-360/C360_2011_brochure_FINAL.pdf.

Index

Printed by Printforce, the Netherlands